The ProR Approach: Traceability of Requirements and System Descriptions

Inaugural-Dissertation

zur Erlangung des Doktorgrades
der Mathematisch-Naturwissenschaftlichen Fakultät
der Heinrich-Heine-Universität Düsseldorf

vorgelegt von

Michael Jastram
aus Reinbek

Düsseldorf, Mai 2012

aus dem Institut für Informatik
der Heinrich-Heine-Universität Düsseldorf
B61

Gedruckt mit der Genehmigung der
Mathematisch-Naturwissenschaftlichen Fakultät
der Heinrich-Heine-Universität Düsseldorf

Referent: Prof. Dr. Michael Leuschel
Korreferent: Prof. Dr. Peter Fromm

Tag der mündlichen Prüfung: 26. Juni 2012

ISBN-13: 978-1478220060
ISBN-10: 1478220066

Es ist nicht genug zu wissen –
man muss auch anwenden.

Es ist nicht genug zu wollen –
man muss auch tun.

Johann Wolfgang von Goethe

Abstract

Creating a system description of high quality is still a challenging problem in the field of requirements engineering. Creating a formal system description addresses some issues. However, the relationship of the formal model to the user requirements is rarely clear, or documented satisfactorily.

This work presents the ProR approach, an approach for the creation of a consistent system description from an initial set of requirements. The resulting system description is a mixture of formal and informal artefacts. Formal and informal reasoning is employed to aid in the process. To achieve this, the artefacts must be connected by traces to support formal and informal reasoning, so that conclusions about the system description can be drawn.

The ProR approach enables the incremental creation of the system description, alternating between modelling (both formal and informal) and validation. During this process, the necessary traceability for reasoning about the system description is established. The formal model employs refinement for further structuring of large and complex system descriptions. The development of the ProR approach is the first contribution of this work.

This work also presents ProR, a tool platform for requirements engineering, that supports the ProR approach. ProR has been integrated with Rodin, a tool for Event-B modelling, to provide a number of features that allow the ProR approach to scale.

The core features of ProR are independent from the ProR approach. The data model of ProR builds on the international ReqIF standard, which provides interoperability with industrial tools for requirements engineering. The development of ProR created enough interest to justify the creation of the Requirements Modeling Framework (RMF), a new Eclipse Foundation project, which is the open source host for ProR. RMF attracted an active community, and ProR development continues. The development of ProR is the second contribution of this work.

This work is accompanied by a case study of a traffic light system, which demonstrates the application of both the ProR approach and ProR.

Zusammenfassung

Eine Systembeschreibung hoher Qualität zu erstellen ist nach wie vor ein große Herausforderung im Bereich Anforderungsmanagement. Einige der Schwierigkeiten können mit formalen Systembeschreibungen verbessert werden. Allerdings ist der Zusammenhang zwischen dem formalen Model und den Nutzeranforderungen selten klar oder zufriedenstellend dokumentiert.

In dieser Arbeit wird er ProR-Ansatz vorgestellt, der die Erstellung einer konsistenten Systembeschreibung ermöglicht, die aus einem initialen Satz von Anforderungen entwickelt wird. Die sich daraus ergebende Systembeschreibung besteht aus einer Mischung von formalen und formlosen Artefakten. Formale und formlose Beweisführung unterstützen den Prozess. Um dies zu ermöglichen, muss es eine Nachverfolgbarkeit der Artefakte zur formalen und formlosen Argumentation geben. Diese Nachverfolgbarkeit unterstützt und ermöglicht es, Aussagen über die Systembeschreibung zu machen.

Der ProR-Ansatz ermöglicht den inkremellen Aufbau der Systembeschreibung, indem abwechselnd modelliert (formal und formlos) und validiert wird. Dabei wird die für die Argumentation notwendige Nachverfolgbarkeit aufgebaut. Das Formale Modell kann Verfeinerung einsetzen, um große und komplexe Systembeschreibungen zu strukturieren. Die Entwicklung des ProR-Ansatzes ist der erste Beitrag dieser Arbeit.

In dieser Arbeit wird ProR vorgestellt, eine Werkzeugplattform fürs Anforderungsmanagement, die den ProR-Ansatz unterstützt. ProR wurde mit Rodin integriert, einem Werkzeug für die Event-B-Modellierung. Mit Hilfe dieser Werkzeugplattform kann der ProR-Ansatz skalieren.

Der Kern von ProR ist unabhängig von dem ProR-Ansatz. Das ProR zugrunde liegende Datenmodell basiert auf dem internationalen ReqIF-Standard, welcher Interoperabilität mit industriellen Werkzeugen im Anforderungsmanagement ermöglicht. Die Entwicklung von ProR hat genug Interesse geweckt, um die Gründung des Requirements Modeling

Frameworks (RMF) zu rechtfertigen, zu dem ProR nun gehört. RMF ist ein Eclipse Foundation-Projekt, und ProR somit Open Source. Die Entwicklung von ProR ist der zweite Beitrag dieser Arbeit.

Weiterhin enthält diese Arbeit eine Fallstudie eines Ampelsystems, welches die Anwendung des ProR-Ansatzes und von ProR demonstriert.

Contents

Chapter 1

Introduction

This work is concerned with the creation of a system description from an initial set of requirements, consisting of *artefacts*. The resulting system description provides traceability between its artefacts, which may be informal or formal. This work describes both the structure of the system description, as well as the process of creating it. This is called the ProR approach. The approach is supported by a tool called ProR. This chapter sets the stage and introduces the concepts that will be used throughout this work.

In the following, only a few selected literature citations on key papers. A much more elaborate list of related work is found in Chapter 2.

1.1 Specifying Systems

Everything is built twice: First an idea forms in the mind, then the idea is realised. This is true from the smallest to the biggest projects, from hanging up a picture to building a space rocket. In the case of the space rocket, there would be a number of intermediate steps to account for the complexity and scale of the task at hand.

The number of intermediate steps and types of documentation depends on the size of the project, how critical it is, how many people are involved and many other factors. But to give just one concrete number, Capers Jones reports that a software system with one million lines of code requires an average of 69 kinds of documentation [McConnell, 2004]. Requirements and specification are just two kinds of documentation — from this it should be clear that this work only covers a small aspect of the development process of a big project.

15

Nevertheless, requirements and specification are artefacts that are so important that they play part in all but the smallest projects. In fact, I'd argue that they are the most important artefacts on the way from start to finish. Figure 1.1 depicts the major milestones that lead from an idea (goal) to the actual thing (implementation), with requirements and specification as the intermediate steps. This is illustrated in the following:

Figure 1.1: The four major milestones that lead from goal to implementation

Every project has a *goal*, let it be as simple as "decorating the wall" or as ambitious as "landing a man on the Moon and returning him safely to the Earth" [Kennedy, 1961]. A goal typically says nothing about the "how" ("How do I achieve this?"), but the "what" ("What is it that I want to achieve?"). A goal is typically very simple and high-level. This does not necessarily mean that it is not precise or quantifiable (consider Kennedy's speech: the goal is certainly precise, and definitely measurable).

A *requirement* puts the goal into the context of the world. A requirement for hanging a picture on a wall is that it stays there, which in turn has to take the picture's properties into account: The requirements for hanging a 10 × 15 cm photograph certainly differ from hanging an art work that is cast from plaster, measures 1 × 1.5 m and weighs over 200 kg. Note that the requirement still should not indicate *how* the picture is mounted — one nail, two nails, screws, glue — because a good requirement does not provide a solution, but precisely describes the problem. (What constitutes a "good" requirement is discussed in Section 2.2.)

For large projects, like going to the moon, it is not practical to go directly from goal to requirements. The goal is typically broken down in sub-goals, an overall architecture is established that allows partitioning of the tasks at hand, etc. In addition, there is a lot of overhead that does not directly contribute to the development, but that is crucial nevertheless. This includes artefacts for sub-disciplines like project management, testing, supply chain management, and many other areas of interest.

It is the *specification* that provide a solution to the problem. This is the place that describes that a nail shall be used to put up the photo, and where to put it. It is dangerous to look for solutions sooner than at this point, because it is easy to miss an important requirement or something crucial regarding the context.

For big projects the step from requirements to specification is challenging, because so many requirements have to be taken into account, and interdependencies can easily be missed. Therefore, big projects employ various techniques to cope with this (see Section 2.1).

The last step is the *implementation* of the system. This can be done in five minutes, using a hammer and a nail. In this case, few would bother to write down requirements or specifications. In fact, in a simple scenario like hanging up a picture, one would probably not even consciously notice that there are implicit requirements and a specification.

Or it may take ten years. Sending men to the moon and bringing them back safely took massive resources and required extensive collaboration. The requirements and specification are crucial tools that enable collaboration in the first place — nobody would attempt such a project without them (and a good measure of bureaucracy thrown in as well).

1.1.1 Iterative Development

Let's get back to Figure 1.1: it may not clear what the meaning of the arrows is. It could be "time" or a "results in" relationship. Either way, the picture implies a linear relationship akin to the so-called "waterfall approach" [Royce, 1970], where no phase of the development process is started until the preceding is completed, while never returning to completed phases. In fact, Royce uses the waterfall approach in the cited paper as an example of a flawed development process.

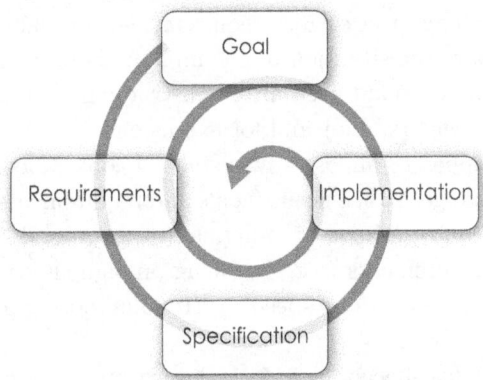

Figure 1.2: The construction process

In reality, for all but the simplest systems, the development is iterative, as suggested in Figure 1.2. Even if the goal is clear (not always the case), the requirements and domain properties still have to be elicited, which, depending on the domain and the stakeholders, is not trivial. Chances are that some are missed or incomplete. During the creation of the specification, problems with the requirements can be uncovered and should be fixed right away. Further, partitioning of the system may result in different progress of different parts of the system.

In Figure 1.2, the implementation is part of the iterative development process. This is actually common in some environments. For instance, the Scrum method expects runnable code at the end of each iteration, which typically lasts from two weeks to two months. On the other hand, the development of embedded controllers typically only produces an implementation towards the end of the development. Approaches to system development are discussed further in Section 2.1.

1.1.2   Stale artefacts

There are a number of reasons why stale artefacts are a bad thing, both during development and after completion. Figure 1.2 already gives a hint why stale artefacts during the development are problematic: Any artefact may be revisited an arbitrary number of times. For example, encountering a requirement that contains outdated or incorrect information could result in the addition of an incorrect specification element, or even to breaking something in the specification that was correct before.

One could think that the artefacts do not matter after the development is completed, but that is not true. In the maintenance phase, it is highly advantageous to have the correct requirements, specification and other artefacts. Compare the situation to a completed house: surely one would keep the blueprints around even after construction. They are invaluable both for service (repairs, etc.) and for extensions.

The same applies to almost any system. People's needs change, and that results in changes to the needs (goals and requirements) of the system under development. If bugs or other issues are found, they are often related to the requirements (e.g. by being ambiguous or contradicting). Maintaining the artefacts can speed up troubleshooting and adding new features significantly.

Especially in fast-paced, non-critical software systems, this maintenance is often not performed. Of course, life will still go on, but the neglect comes at a price. How high depends on the precise circumstances. For instance, a typical statement of people who prefer not to document is "the code is the documentation". And this can actually be true. Well written, well commented code should certainly be considered to be part of the documentation. Likewise, the test code (if it exists) is also part of the documentation, and running the tests regularly ensures that this part of the documentation is up to date (at least in respect to the code). And if the code is based on a well known architecture or on a well documented framework, that part of the documentation can be slim. An example is the Ruby on Rails framework [Hansson et al., 2012], which provides a certain part of the architecture and that consequently does not require additional documentation (except in those places where the system deviates from the framework's conventions).

But again, in reality one is often confronted with outdated artefacts. The reason is that maintaining artefacts manually is hard, and most of the maintenance is done manually. In fact, even without the interdependency between artefacts things are hard. Consider a requirements document of one hundred pages, which is not unusual, written in natural language. Who can be sure that there is not a contradiction between a statement on page 5 and on page 95? Who has the confidence to detect such a contradiction with certainty?

Last, keeping and maintaining artefacts must not become an end in itself. There will be benefits of up to date artefacts, but there is a cost to, the cost of maintenance. If the cost is too high, it may be better to dispose the artefact after it has been used, rather then keeping a stale version around. Of course, understanding the cost and benefits is not easy, and one tends to underestimate long term benefits and overestimate short-term costs. Further, the cost-benefit-ratio can sometimes be shifted by using a new approach or a different tool for the job.

1.1.3 Structuring the System Description

So far, the discussion of the artefacts from Figure 1.2 was rather casually. But quite a bit of research went into understanding this relationship better. Gunter, Jackson and Zave [Gunter et al., 2000] developed WRSPM as a *reference model* for requirements and specifications. A reference model is attractive for discussion, as it draws on what is already understood about requirements and specifications, while being general enough to be flexible. There are a number of concrete approaches that fit nicely into the WRSPM reference model.

This thesis introduces a modified WRSPM model (Section 3.2.2). The central artefacts of that model are:

Domain Properties (W) describes how the world is expected to behave.

Functional Requirements (R) describe how the world should behave.

Non-Functional Requirements (N) describe quality properties of the system to be build.

Specifications (S) bridge the world and the system.

Design Decisions (D) justify why the system was specified in a certain way.

Domain properties W and requirements R and N are typically found in the requirements, according to Figure 1.2. There is a fundamental difference between them: Requirements describe how the world *should behave*, and the system is responsible for this. Domain properties describes how the world is *expected to behave*, and the functioning of the system depends on the domain knowledge holding.

Note that WRSPM does not have the concept of a goal. According to WRSPM, a goal is merely a high-level requirement.

The specification in Figure 1.2 corresponds to S and describes how the requirements are to be realised, in the context of the domain.

The implementation from Figure 1.2 corresponds to the program P, which is another WRSPM artefact. Implementation is further discussed in Section 3.2.2.

The reference model also defines *phenomena*, which act as the vocabulary to formulate the artefacts. Phenomena are terms that typically designate states, events, and individuals (see Section 3.2.1) There are different types of phenomena based on their visibility. For instance, there may be phenomena that the machine is not aware of. Consider a thermostat: the controller is not aware of the temperature[1], but only of a voltage at one of its inputs.

The reference model can be applied to any requirement or specification, no matter whether they use natural language or a formalism. Once applied, more formal reasoning about the specification is possible, based on the classification of artefacts and phenomena and their relationship to each other.

---

[1]To be precise, whether the controller is aware of the temperature or not depends on where the line is drawn between system and environment. In this simple example, the sensor is not part of the system (the controller).

Stakeholders rarely explicitly distinguish between requirements, domain knowledge, or even specification elements and implementation details. More technical details of the ProR approach are covered in Section 3.2.1.

1.2 Traceability

Traceability refers to the relationships between and within the artefacts (i.e., elements of W, R, N, S, or D). In this work, the relationship of artefacts and phenomena is clearly defined (Section 3.2.2). There may be more relationship outside the design description (i.e. tests, project management, etc.). These are plentiful and exist implicitly. In the following, traceability of system description artefacts is discussed in more general terms.

The arrows in Figure 1.1 suggest a possible traceability between the elements of the shown "milestones". This could be an "is realised by" traceability, indicating that the milestone on the left (e.g. goal) is realised by the elements from the milestone to its right (e.g. the requirements).

There are many more relationships. Consider a textual requirements document. The order of the requirements represents a relationship between the requirements to each other. Many requirements documents have a glossary: There is a relationship between a glossary term and the use of the term in the requirements. There is (or should be) a causality between the requirements and the specification: Any feature in the specification should be justified by a requirement.

The implicit traceability can be made explicit. But by doing so, those traces become themselves artefacts that must be maintained. Therefore, the benefits and costs of making traces explicit must be weighed carefully — as with some artefacts, the cost of stale traces may be higher than the cost of no explicit traces.

Making traces explicit can in itself provide useful information. Consider the "is realised by" relationship between requirements and specification. Such a relationship would immediately identify those requirements that are not specified yet, namely those requirements that have no outgoing traces. Such a requirement can then be inspected and the specification extended to realise it. After the specification has been extended, a new trace is created, marking the requirement as realised.

While this approach works in principle, there are at least two problems with it. First, which elements will be traced? It would be nice if there was a one-to-one relationship between requirements and specification elements, but this is true only for the simplest toy examples. In practice, this is an n-to-m relationship, and sometimes one end of the trace can be elusive: Consider quality requirements that apply to the system as a whole: non-functional requirements like performance, user experience, responsiveness, etc. are typically the result of the system as a whole, and are difficult to trace to individual specification elements.

Maintenance is the second issue. Creating a trace correctly is one thing, but keeping it updated is quite another. Consider again the "realised by" relationship. All incoming traces would have to be verified to make sure that the specification element still, in fact, realises all requirements that it traces. But this works only if all traces have been created in the first place. And when more corrections have to be done during this verification (both on requirements and specification), it may trigger another wave of verifications. Tool support can help to mark traces for verification — but how much this helps depends on the completeness and correctness of the traceability.

The ease of traceability depends, amongst other things, on the structure and quality of the artefacts. For instance, one quality criteria for good requirements is the absence of redundancy. Not having redundancy also eases traceability. Further, there are many ways to structure the artefacts. A good structure can make traceability significantly easier. The structure depends on notation and approach. The approach guides the artefacts towards a certain structure, while the notation constrains how easy or difficult it is to express something. Some notations require a certain approach and may also push the artefacts in a certain structure. This is good if the notation is well-suited for the problem at hand, but it can be counter-productive if this is not the case. Just imagine drawing the blueprint of a house with UML, or to document an enterprise-system with a mechanical drawing. Other notations are highly expressive, like natural language. But the downside in this case is that the notation provides no guidance, and can be ambiguous.

The **ProR approach** addresses these issues. It imposes a classification scheme on the system description artefacts and phenomena. It then introduces a small number of well-defined traceability relationships, including "realises", "justifies" and "uses" (see Section 3.2 for the details). Maintenance is addressed by providing a mechanism for marking traces as *suspect*, and thereby marking them for re-validation.

This discussion already gives a glimpse of the potential cost of establishing and maintaining a traceability. Some of this cost can be eased by tool support and automation, as discussed in Chapter 4.

An analysis of current research in the area of traceability is presented in Section 2.3.

1.3 Modelling

The structuring of artefacts as presented so far provides the foundation for further modelling. A model is anything used in any way to represent certain aspects of something else, in this case of the system to be build. Many modelling languages exist, all with their respective advantages and disadvantages. Modelling can also be applied on various levels of the development process — for goals, requirements, the specification and even for the implementation.

Models typically constrain the structure of the model by their model elements or their meta-model. Consider context diagrams, a notation that can be used in the requirements elicitation process and that forces us to define the boundary of the system and to identify the actors that can interact with it (Figure 1.3). Using a context diagram in the elicitation process will leave its traces in the structure of the requirements (i.e. by systematically enumerating all actors and how they interact with the system).

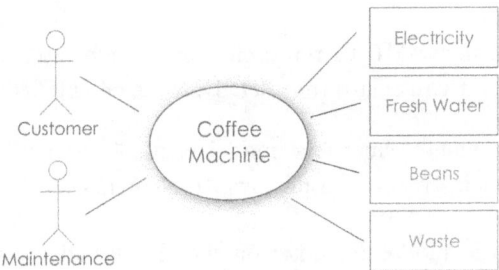

Figure 1.3: An informal context diagram for a coffee machine

1.3.1 Semi-Formal Modelling with SysML

Context diagrams are very informal. SysML[2] is a notation that is sometimes referred to as "semi-formal", in that it is much more strict than context diagrams, but not as rigorous as a mathematical notation.

---

[2]SysML is a variation of the better-known UML. In contrast to UML, it contains the model element "requirement".

UML and SysML contain some modelling rules and supports refinement activities, to a degree. However, systematic verification is not possible (with respect to consistency), merely a syntactic check of the connectivity of elements.

SysML also contains context diagrams (called block definition diagrams). These are "more formal" than regular context diagrams, in the sense that they have well-defined traceability relationships. Specifically, SysML contains requirements diagrams holding requirements, which can have five types of traces that allow to establish relationships between requirements and other SysML model elements. The relationships are simple, which is one of the appeals of the method in industry. Requirements in SysML are essentially just boxes with text in it. SysML provides the following trace types:

Containment SysML allows requirement containment, with respect to another requirement or name space (e.g. block or package).

DeriveReqt Indicates that one requirement is derived from another requirement (but not other SysML elements).

Satisfy Describes how a design or implementation model satisfies one or more requirements.

Verify Is used in SysML to represent verification, etc. by a test case. Therefore, it can relate to other SysML elements as well.

Refine Requirements can refine model elements and vice versa, which may result in traces to other SysML elements.

Note that the modelling notation merely provides these model elements, but essentially it's just boxes with attributes (which are text), connected by named traces. This may be useful for structuring requirements and the specification, but it does not allow formal reasoning.

1.3.2   Formal Modelling with Event-B

In contrast to a semi-formal notation like SysML, a formal notation will allow rigorous reasoning and is typically based on a mathematical notation.

Formal methods are a particular kind of mathematically-based techniques for the specification, development and verification of software and hardware systems. Event-B is a formal method for system-level modelling and analysis. Key features of Event-B are the use of set theory as a modelling notation, the use of refinement to represent systems at different abstraction levels and the use of mathematical proof to verify consistency between refinement levels.

Event-B can be used for writing a subset of the specification (see Figure 1.2) by representing certain artefacts formally, according to 1.1.3. Doing so can have certain advantages. For instance, a requirement expressed as an invariant can be proven to never being violated — at least, if the model captures the problem correctly.

While it is possible to express requirements and domain properties exclusively in Event-, users may reject this. It is just not comprehensible to the majority of the stakeholders[3]. Event-B can express desired properties on a high level in a concise manner and then use refinement to add more and more details, even so far that constructing an implementation is merely a mechanical translation process.

Not all artefacts can be captured concisely with Event-B. For those, either other formalisms can be used, or they can be omitted from the model altogether. Artefacts that are omitted must be considered at a different point of the development process.

Event-B does not have the notion of requirements or traceability. The approach presented in this work aims to provide this: It combines Event-B with other elements to create a traceability between requirements and specification. This approach does not assume that Event-B is well-suited for providing a specification for all requirements, as opposed to some other approaches. It allows the selection of those requirements that should be modelled formally and those that are specified otherwise.

More details regarding Event-B and other formal methods are found in Section 2.7.

1.4 Systems Development

Humans have been building systems for centuries, and are actually pretty good at it: trains keep running, planes are astonishing reliable. They are very careful to not exclude approaches that are proven to work. They expect new work to complement existing approaches, not to replace them.

Today's system development approaches are typically unique to the organisations where they are practises — in fact, sometimes they are unique

---

[3]There may, of course, be situations where all stakeholders are fluent in predicate logic. In those cases use Event-B all the way, by all means! But this is the exception, not the rule.

to individual departments or even projects. This has been recognised a
long time ago, and today's approaches are designed to be tailored. A well-
known approach in software development in the Rational Unified Process,
another one in systems development is the V-model, which is depicted in
Figure 1.4.

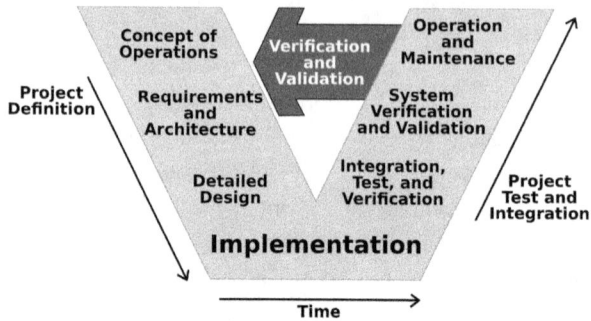

Figure 1.4: V-Model (Image Source: Wikipedia)

The V-model does not prescribe how the individual steps are realised
— it is up to the user of the process to fill it with live. Nevertheless,
the arrow ("Verification and Validation") implies a traceability. It also
implies that the validation process is done after the project definition and
implementation stage. This is a significant difference to the approach
presented here: It suggests to do a significant amount of validation during
the project definition stage.

The V-model emphasises requirements-driven design and testing. All
design elements and acceptance tests must be traceable to one or more
system requirements and every requirement must be addressed by at least
one design element and acceptance test. Such rigour ensures nothing
is done unnecessarily and everything that is necessary is accomplished
[K. Forsberg and Cotterman, 2005]. However, it is one thing to postulate
this and another to realise this. The details are left to the implementer of
the V-model.

The approach presented in this work is no full-blown process. It does
not cover all aspects of systems development, but instead offers concrete
solutions for certain aspects of the system development process. There are
many elements of the V-model, for instance, that could be nicely covered
with this approach. An existing development process (based on the V-
model) can be modified in small steps to include more and more elements
of this approach.

1.5   A List of Original Contributions

This research makes a significant contribution in providing a comprehensive theory of tracing formal and informal elements of a system description. A tangible result of this work is the ProR platform for requirements engineering, an open source tool which filled a vacuum in the Eclipse ecosystem with respect to system engineering. ProR development continues with an active community that consists of academic and industrial contributors.

The original contributions made by this research are listed below.

1.5.1   The ProR Approach

The papers listed here are concerned with the development of the ProR approach (3) and may also mention tool support. Some also contain work relating to the case study (Chapter 5).

Mixing Formal and Informal Model Elements for Tracing Requirements [Jastram et al., 2011]

This research represents the foundation of the ProR approach. It brings together the structuring of informal specification artefacts using the WRSPM approach and formal state-based modelling, including a theory to traceability. My contribution concerns mainly the extension of the WRSPM reference model and ProR tool support, as well as the case study.

An Approach of Requirements Tracing in Formal Refinement [Jastram et al., 2010]

This research demonstrates the ProR approach on a case study in which traceability between natural language requirements and an Event-B model is developed. My contribution concerns mainly the extension of the WRSPM reference model and ProR tool support, as well as the case study.

Requirements, Traceability and DSLs in Eclipse with the Requirements Interchange Format (RIF/ReqIF) [Jastram and Graf, 2011d]

While this research is focused on tool support, it also demonstrates how two different approaches to traceability were realised, using the ProR platform for requirements engineering. My contribution concerns the Deploy-related research results and ProR tool support.

Strukturierung von Anforderungen für eine enge Integration mit Modellen [Jastram, 2012a]

This work was presented in an industrial context and demonstrates the applicability of the ProR approach in practice.

1.5.2   Tool Support: ProR and RMF

The papers listed here are primarily concerned with tool support (Chapter 4), consisting of a mix of academic, semi-academic and industrial publications.

ReqIF – the new Requirements Standard and its Open Source Implementation Eclipse RMF [Jastram and Graf, 2012]

This work represents an up to date status on the Requirements Modeling Framework, the Eclipse project that ProR is part of. It is a confirmation of ProR's success in the requirements engineering community. My contribution concerns the GUI-related aspects of RMF.

Requirements Modeling Framework [Jastram and Graf, 2011c]

Like the previous paper, this is a recent publication in an Eclipse-focused industry magazine, emphasising the practical relevance of the ProR tool and project hosting it, the Eclipse Requirements Modeling Framework.

ProR, an Open Source Platform for Requirements Engineering based on RIF [Jastram, 2010]

This work presents ProR at an early stage, before it became an Eclipse Foundation project. It provides a benchmark on the progress of tool support, with respect to the two previous publications.

ProR – Eine Softwareplattform für Requirements Engineering [Jastram, 2011]

This is another German-language publication that presents ProR at an early stage.

Requirement Traceability in Topcased with the Requirements Interchange Format (RIF/ReqIF) [Jastram and Graf, 2011b]

This work outlines the potential of ProR to be integrated with other Eclipse-based offerings, in this case the Topcased tool for UML and SysML modelling. My contribution concerns the GUI-related aspects of ProR and the Deploy-related research results.

ReqIF in der Open Source: Das Eclipse Requirements Modeling Framework (RMF) [Jastram and Brökens, 2012]

This work was presented in an industrial context and demonstrates the applicability of ProR and RMF in practice. My contribution concerns the GUI-related aspects of RMF.

1.5.3 Umpublished, Accepted Work

Work on both, the ProR approach and the tool continues. The work listed here has been accepted for publication.

A Method and Tool for Tracing Requirements into Specifications [Hallerstede et al., 2012]

This paper builds heavily both on [Jastram et al., 2011] and this work, but expands on the theory with respect to the Event-B formalism. It also provides a case study that is unrelated to the one presented in this thesis. My contribution concerns mainly the extension of the WRSPM reference model, the process description and ProR tool support.

Managing Requirements Knowledge Book [Jastram, 2012b]

This book focuses on potentials and benefits of lightweight knowledge management techniques applied to requirements engineering. My contribution is the chapter "The Eclipse Requirements Modeling Framework".

ReqIF: Seamless Requirements Interchange Format between Business Partners [Jastram and Ebert, 2012]

This paper looks at the ReqIF standard and the RMF-based tool chain from the point of view of project risks and product problems. My contribution concerns all RMF-related aspects of this paper.

1.6 Summary

This chapter presented the basic steps and artefacts in systems development — no matter whether something small or gigantic is to be built. For all but the smallest projects, artefacts can get stale quickly, and reasoning can be challenging. Formal models can make reasoning easier, but are not always comprehensible to all stakeholders. A functional traceability can help to identify and remedy stale artefacts.

In the following pages, a new approach to formal traceability is presented. This approach is intended to be combined with existing systems development processes. It allows for only a part of the system to be modelled formally, while leaving the rest to traditional methods; it provides a traceability between requirements and formal specification elements. And last, a tool platform is presented that is based on industrial standards, as well as a proof of concept implementation.

1.7   Acknowledgements

This work would not have been possible, had I not gotten support from a diverse group of people. I would like to thank Prof. Michael Leuschel for giving me the opportunity and freedom to develop my ideas as I saw fit. Further, the STUPS research group was extremely helpful and full of advise, and always willing to listen, in particular Jens Bendisposto, Daniel Plagge and Stefan Hallerstede. I'd also like to thank Peter Fromm who agreed to review this work on a tight schedule.

I am grateful for the Deploy project, which provided some of my funding. But beyond funds, many people in this group gave me invaluable advice and provided guidance. For this I'd like to thank foremost Cliff Jones, but also Michael Butler, Mike Poppleton, Colin Snook and Rainer Gmehlich and Felix Lösch.

ProR would never be as mature as it is now, had we not joined forces with the Verde project. In particular, I'd like to thank Andreas Graf and Nirmal Sasidharan for their collaboration. Without them, Eclipse RMF would probably not exist today. ProR takes advantage of the OpenSource AgileGrid control, and I'd like to thank its creator Sihong Zhu.

And last, I would like to thank my wive Maha for keeping my back free, so I could work on this thesis, as well as my parents, who always encouraged me to advance my knowledge.

1.8   Formal Mind

One byproduct of this work is the founding of Formal Mind GmbH[4] by Prof. Michael Leuschel, Jens Bendisposto, Daniel Plagge and myself.

---

[4]http://formalmind.com

Formal Mind is active in the area of systems engineering and bases its products and services on the work described in this thesis, as well as ProB [Leuschel and Butler, 2003]. Stimulating the foundation of new companies was one of the objectives of the Deploy project [EU FP7 Project, 2012].

Chapter 2

Literature and Related Work

This thesis is concerned with traceability of system descriptions. This chapter provides an overview of the existing work in the areas of requirements, formal models and traceability. Part of this work is a tool platform for requirements management, therefore there will be an overview of the tool landscape.

The activities involved in the approach presented here fall in the field of system development, which is covered in the next section in order to put the work presented here into context.

2.1 Systems Development

In Section 1.4, the concept of systems development and the V-model (Figure 1.4) were introduced. The V-model was published first in 1986 and was developed for the federal government of Germany. It has been revised several times, the last one being the release of the V-model XT in 2005 [Broy and Rausch, 2005].

Another detailed development method is the Rational Unified Process (RUP) [Kruchten, 2004] and its "cousin" the Open Unified Process (OpenUP). OpenUP [Balduino, 2007] is a revision of the iterative Rational Unified Process for software development process that is minimal, complete, and extensible. The OpenUP can be browsed online[1].

V-model and RUP fall in the category of "heavyweight" development

---

[1] http://epf.eclipse.org/wikis/openup/

methods, in contrast to "lightweight" (or agile) methods like extreme programming (XP) [Beck, 2001] or Scrum [Schwaber, 2004]. There are many more, and a comparison can be found in [Awad, 2005].

In most development methods, there is the concept of requirements and the concept of a specification. Even light approaches like XP capture requirements, although typically in a very practical format, like user stories [Cohn, 2004]. They also have the concept of a specification, although the interface of a class may already be sufficient as a specification with these approaches.

Contrast this with OpenUP, which lists five different requirements artefacts (Glossary, System-Wide Requirements, Use Case, Use-Case Model and Vision), and various artefacts regarding the specification.

The standard IEEE 830-1998 [IEEE, 1997] is concerned with software requirements specifications (SRS) and is heavily used and cited in industry. It is concerned both with the end result (the SRS), as well as the process of authoring it. It provides completeness and quality criteria for SRS that concern individual requirements as well as the SRS as a whole. In that regard, it provides sample outlines that can be used as a starting point of creating an SRS. The standard is independent of a notation or a specific tool. Therefore, even though it is targeted at industrial users who use natural language requirements, it explicitly acknowledges that it could be applied to specialised specification languages and points out the risks associated with such an approach. The standard explicitly doesn't cover aspects like design or project requirements.

Reveal [Praxis, 2003] is an engineering method based on Michael Jackson's "World and the Machine" model, which is applied in industry by the company Altran Praxis[2]. There are some similarities to the approach described in this thesis, including the acknowledgement of requirements that are not part of the formal model. However, Reveal is more of a process description of the overall requirements engineering process. It could be quite attractive to combine the Reveal process with the approach described here.

2.2   Requirements, Domain Properties and Specifications

Requirements and specifications are central in this work. In [Zave, 1997], the authors provide an overview of research efforts in requirements

---

[2]http://www.altran-praxis.com/reveal.aspx

engineering by providing a classification scheme. It is intended to provide an overview and a coherent framework for further study.

Section 2.4 contains a well-defined definition of the terms "requirement" and "specification", according to WRSPM. But these terms also have a slightly different meaning in the context of requirements engineering and common language, which is covered in the following section.

According to [Project Management Institute, 2008], a requirement is:

(1) a condition or capability needed by a user to solve a problem or achieve an objective;

(2) a condition or capability that must be met or possessed by a system, system component, product, or service to satisfy an agreement, standard, specification, or other formally imposed documents;

(3) a documented representation of a condition or capability as in (1) or (2);

(4) a condition or capability that must be met or possessed by a system, product, service, result, or component to satisfy a contract, standard, specification, or other formally imposed document. Requirements include the quantified and documented needs, wants, and expectations of the sponsor, customer, and other stakeholders.

This definition is also referenced by various standards, e.g. ISO/IEC/IEEE 24765, which is widely used in industry.

The approach described in this work uses natural language requirements as the starting point. While requirements can be stored in forms other than natural language, it is the most natural way for stakeholders to express their perception of the model [Ambriola and Gervasi, 1997]. Natural language requirements can also be processed (semi-)automatically [Goldin and Berry, 1997], which is outside the scope of this work.

The quality of requirements is a big concern, as all subsequent artifacts in the development process depend on them. The quality of requirements can be improved by using a catalogue of criteria, as described in [Hood and Wiebel, 2005], or by analysing linguistic properties [Fabbrini et al., 1998]

Various approaches exist to categorise requirements. It is common to distinguish between functional and non-functional requirements, where a functional requirement describes a system in terms of inputs and expected outputs (as well as exceptions) [Rupp, 2007]. A nonfunctional requirement does not, well, relate to functionality. Instead, it relates to attributes like usability, performance, reliability, etc.

Pohl [Pohl, 2007] prefers the separation into functional and quality requirements. The motivation is the "abuse" of non-functional requirements
in practice, which he claims are often under-specified functional requirements. These can typically be broken down into more detailed functional
requirements and quality requirements.

The distinction between functional requirements and "others" is in so
far important in this work, as only functional requirements are typically
modelled formally. In fact, depending on the formalism employed, even
some functional requirements may be difficult to model. There is a lot
of material regarding this issue. In [Chung and do Prado Leite, 2009],
the authors provide a good introduction into the matter, while there are
a number of methods that provide guidance in designing a system's non-
functional requirements, e.g. [Gross and Yu, 2001]. Some approaches take
one step further back and introduce goals, as we will see in the next section.

A domain property is something that is *assumed* to be always true. In
practice, domain properties are often recorded together with requirements.

For non-experienced users it is easy to confuse requirements and
domain properties, as they both are expected to hold for the final system.
But the main difference stems from the underlying reason for them to
hold:

- The domain properties hold irrespective of the system that was built.
  If they do not hold due to external influences, the system cannot be
  expected to continue to work as designed.

- The requirements hold because the system was designed that way,
  but only under the assumption of the domain properties holding.

It is not unusual to omit some domain properties, because they are
"obvious". This can be the source of severe problems. A prominent
example is the crash of the Ariane 5 rocket on June 4th, 1996 [Consulting
and Ninomiya, 1997]. The Ariane 5 software reused the specifications
from the Ariane 4. Due to the rocket's greater acceleration, a data
conversion from a 64-bit floating point to 16-bit signed integer value caused
an arithmetic overflow, ultimately destroying the rocket. The underlying
problem was the omission to record the maximum acceleration as a domain
property.

Many more formal approaches recognise the need to record domain
properties, including WRSPM (Section 2.4) and Problem Frames Section 2.5. Another approach that makes this explicit is the Rely-Guarantee
approach, which explicitly lists the properties that a system component
relies upon [Coleman and Jones, 2007].

Informally, a specification is the name for a system description that consists of requirements. DeMarco [DeMarco, 1979] provides a classic description of the specification and its creation process. This is quite different from the definition in Section 2.4, where the term specification refers to all the information that a developer needs to build a system that fulfils the given requirements.

The purpose of the specification is to describe accurately how the system to be designed fits into the real world. In [Jones et al., 2007], the authors argue that the task of "fixing" the specification into the external physical world can be more challenging than the development itself.

2.3 Requirements Traceability

Requirements traceability is defined in the International Institute of Business Analysis Body of Knowledge V2.0 as, "The ability to identify and document the lineage of each requirement, including its derivation (backward traceability), its allocation (forward traceability), and its relationship to other requirements." [IIBA, 2009]. Another definition for traceability is the "discernible association among two or more logical entities, such as requirements, system elements, verifications, or tasks" [IEEE, 2010]. It is a difficult problem [Bjørner, 2008, Gotel and Finkelstein, 1994, Jastram et al., 2010].

Abrial [Abrial, 2006] recognises the problem of the transition from informal user requirements to a formal specification. He suggests to construct a formal model for the user requirements, but acknowledges that such a model would still require informal requirements to get started. He covers this approach in [Abrial, 2010].

Rather than creating a traceability from natural language, it is possible to create different kinds of models. This approach is used by the UML-B plug-in [Snook and Butler, 2006], which generates Event-B (Section 2.7) from UML. As the knowledge and understanding of UML is much more widespread than Event-B, it may in some circumstances be acceptable to confront the stakeholders directly with a UML model.

An approach that is concerned mainly with the domain properties is described in [Kaindl, 1997], which creates a traceability between requirements definition and an object model, thereby creating a more complete and better structured definition of the requirements.

A one-to-one traceability between requirements and formal model elements would be desirable, but is elusive. Depending on the formalism, this may be achievable for some requirements. For instance, some safety requirements can be expressed concisely as an invariant, or some timing properties as an LTL expression. This tends to be the exception, rather than the rule, as can be seen in Chapter 5.

An important task of traceability is to support change management. In [Hammad et al., 2009], the authors automatically identify changes that impact code-to-design traceability. This is a useful concept that could be incorporated into the approach described in this thesis.

2.4 WRSPM

The WRSPM reference model [Gunter et al., 2000] was attractive for this work, because it allowed us to discuss the specification in general terms, while still being meaningful in the context of a specific approach like Problem Frames (Section 2.5) , KAOS (Section 2.6) or the functional-documentation model [Parnas and Madey, 1995].

The idea of the WRSPM reference model has been advanced in current research. In [Marincic et al., 2007], the authors introduce a model of formal verification based on non-monotonic refinement that incorporates aspects of WRSPM. Problem Frames (Section 2.5) could be useful for identifying phenomena and for improving the natural language requirements. In [Loesch et al., 2010], the authors show how Event-B and Problem Frames are being applied to an industrial case study. This work drew some inspiration from their efforts, especially with regard to the relation between WRSPM and Event-B.

WRSPM is central to the ideas in this work. The concept of WRSPM is introduced in Section 3.2.1.

2.5 Problem Frames

The Problem Frames [Jackson, 2001] approach aims to properly describe a problem first, before attempting to solve it. It consists of a notation and a method. As realistic problems are too big and complex to handle in just on e step, problem frames allow the structuring as a collection of interacting subproblems. Eventually, subproblems are so small and simple that they fit an existing problem frame, for which the concerns it raises are known.

The Problem Frames notation introduces *problem diagrams*, which extend the notation of context diagrams (see Figure 1.3) that make the problem explicit by showing the requirements in the diagram. The notation of context diagrams is also formalised by distinguishing between machine domain, designed domains and given domains. The notation further introduces *problem frame diagrams* for concisely recording problem frames.

The Problem Frames approach employs the *frame concern*, which "captures the fundamental criterion of successful analysis for problems that fit" a given problem frame. This concept could be applied with the ProR approach for reasoning about informal artefacts (see Section 3.3.2).

The Problem Frames approach is a concrete approach that can be aligned with the WRSPM reference model (Section 2.4. The case study in Chapter 5 uses Problem Frames to improve the initial requirements.

There have been successful attempts in applying Problem Frames and Event-B together. In [Loesch et al., 2010], the authors show how these are being applied to an industrial case study.

2.6 Goal-Oriented Requirements Engineering

Goals can be seen as high-level requirements that capture the objectives of a system. The discipline goal-oriented requirements engineering [Van Lamsweerde et al., 2001] starts with a high-level description of a system (*goal*) and uses them for "eliciting, elaborating, structuring, specifying, analysing, negotiating, documenting, and modifying requirements", where requirements have a well-defined meaning.

A prominent method in this category is KAOS [Darimont et al., 1997]. In KAOS, goals are refined into expectations, domain properties or requirements. Further, a requirement is laced under the responsibility of exactly one agent. Besides the goal model, KAOS provides an object model, a responsibility model and an operation model, providing a complete modelling environment, from high-level goal to formal specification.

Figure 2.1 shows an example from a KAOS model. It shows how a requirement is broken down into sub-requirements and associated with actors that are responsible for fulfilling them. What is shown here is still textual and can be formalised using the KAOS temporal logic notation.

KAOS does not permit requirements to be omitted from the formal model. Instead, it provides so-called "soft-goals" that are broken down into requirements that can still be modelled formally.

While elements from KAOS could in theory be used independently, the approach is designed for everything — either use KAOS all the way or not at all. This can discourage potential users from buying into an approach. I found the approach too constraining to be useful in this work, and suggest a more open and more flexible approach. Nevertheless, if someone would

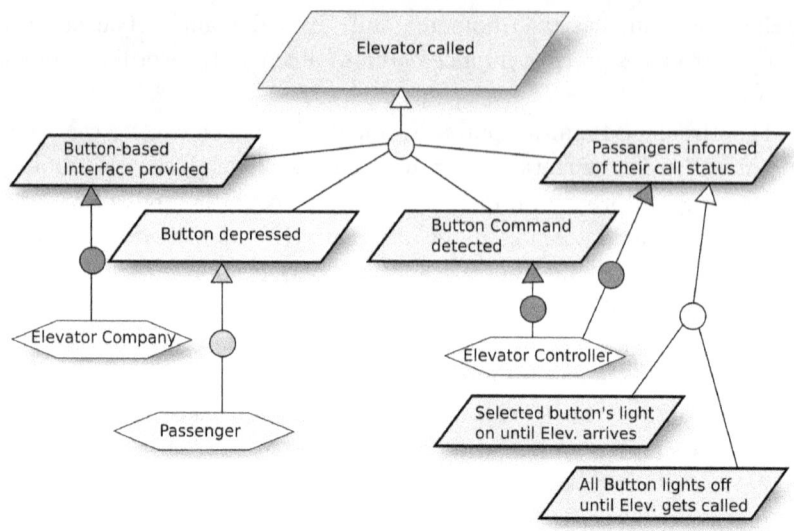

Figure 2.1: Extract from a KAOS model describing an elevator

use KAOS consequently throughout the development from high-level goal to low-level temporal model, this would be very powerful. Reasoning across many layers of the development would be possible, the relationship would disclose dependencies, and the formalism would ensure that the model has certain properties regarding soundness and consistency. How much would depend on the stage at which the model would stop using natural language.

Other methods in this category include NFR [Chung and do Prado Leite, 2009], i* [Yu, 1997] and Tropos. Besides goals, they share common concepts, as described in [DEPLOY Project, 2009]. These include the concept of refinement/contribution between sub-goals: a set of child goals logically entails or, less formally, contributes to the satisfaction of the parent goal — enabling the modelling of system-wide collaborations to reach some goal.

Another shared concept is the conflict/negative contributions between goals, meaning that under some circumstances, two (or more) goals cannot be enforced together. A solution to this dilemma must be found.

Last, these method share the concept that the assignment of a goal to an agent (which is responsible for its realisation) must have the capability to realise it. Such goals are then formally called "requirements".

2.7 Formal Modelling and Event-B

Formal methods are a particular kind of mathematically-based techniques for the specification, development and verification of software and hardware systems. There are many different methods [Wing, 1990, Clarke and Wing, 1996].

Formal methods are used to specify and verify systems. While formal methods do not guarantee correctness, they can greatly increase the understanding of a system and help revealing inconsistencies, ambiguities and incompleteness that might otherwise go undetected [Clarke and Wing, 1996]. A formal specification captures system properties. Depending on the formalism, this can include functional timing properties, performance characteristics or internal structure.

Formal methods can further be distinguished on whether they focus on specifying sequential behaviour (Z [Woodcock and Davies, 1996], VDM [Jones, 1990], Larch [Guttag et al., 1993]) or concurrent behaviour (CSP [Hoare, 1978, Hoare, 2004], CCS [Moller and Tofts, 1990]). Some methods attempt to combine both (RAISE [Nielsen et al., 1989], LOTOS [Brinksma et al., 1995]).

Once a specification is formalised, it can be verified using model checking [Clarke, 1997, Leuschel and Butler, 2003] or theorem proving [Cook, 1971].

This work describes an approach that is independent of a specific formal method. The case study described in Section 5 uses the Event-B formal method [Abrial, 2010], which is described in more detail in Section 3.3.1.

Event-B is considered an evolution of B (also known as classical B [Schneider, 2001]). It is a simpler notation which is easier to learn and use. It comes with tool support in the form of the Rodin Platform [Coleman et al., 2005].

2.8 Traceability between Requirements and Formal Models

There have been attempts to provide traceability between requirements and Event-B models. Abrial applies an informal approach in [Abrial, 2010]. As this approach is applied to the relatively small examples in the book, it is not clear how that approach would scale.

In contrast, [Matoussi et al., 2008] is a first attempt to create traceability between a KAOS model and Event-B, while [Loesch et al., 2010] attempts the same for Problem Frames.

The structure of the requirements determines how well they can be traced, and the Map Requirements Modelling approach [Babar et al., 2007] begins with the structuring of requirements in the elicitation phase and ends with a formal model using the B-Method.

Another approach outlined in [Ball, 2008] describes the development of multi-agent systems using Event-B. The developer is provided with guidance to construct formal Event-B models based on the informal design. This work also includes modelling patterns that provide fault-tolerance for Event-B models of interacting agents. Like the approach described in this work, this is an incremental process.

Chapter 3

The ProR Approach

A challenge in formal modelling is the traceability between requirements and the formal specification. A method for such tracing should permit to deal efficiently with changes to both the requirements and the model. It should not require all requirements to be modelled formally — after all, some requirements may not benefit much from formal modelling.

This chapter introduces the **ProR approach**, an incremental approach that models the system description stepwise by alternating between requirements validation and systems modelling. This results in a scalable approach that supports change management and requirements evolution. The **ProR approach** supports the formal modelling of a subset of the system description. This work expends on what has been presented in [Jastram et al., 2010, Jastram et al., 2011, Hallerstede et al., 2012].

With the **ProR approach**, the initial requirements, provided by the stakeholders, are used to iteratively develop a system description consisting of formal and informal *artefacts*. Artefacts mainly describe requirement items domain properties, specification elements. They are expressed with *phenomena*, which act as vocabulary for the artefacts. Structured this way, reasoning about these artefacts is possible and a central aspect of the **ProR approach**.

Reasoning about the system description can be informal or formal (if all relevant artefacts have been formalised). Even informal reasoning demands firm conclusions with respect to correctness and completeness with respect to the requirements. This is achieved by formal or informal proofs, consisting of validation statements. In order to map validation statements to the system description, a method for *tracing* artefacts is employed.

43

The ProR approach does not require any artefacts to be modelled formally. But it supports the incremental addition of formal artefacts to the system description, which then allows reasoning by formal proof. This takes advantage of the high degree of automation that is possible in formal modelling environments, regarding validation, traceability and formal reasoning. Further, *refinement* can be used for formalising artefacts incrementally.

The ProR approach supports managing change in the system description. It is specifically designed iteratively, as it is expected that the system description changes frequently, particularly in the early development phase. But even after completion, changes tend to continue for maintenance and to provide additional functionality.

The ProR approach is based on the WRSPM reference model [Gunter et al., 2000] (see Section 2.4). It is concerned with classifying artefacts and phenomena in common categories. It specifies soundness conditions that the artefacts must satisfy. This work extends WRSPM by distinguishing functional and non-functional requirements, and by deemphasising implementation details. It then introduces a notion of traces that relates artefacts to proofs (formal and informal), and to react to change by identifying affected artefacts.

The *macroscopic* structure of the system description is orthogonal to the structuring imposed by the ProR approach, and is not central to this work. Macroscopic structuring is concerned with organising large collection of artefacts (in natural language and otherwise), and many approaches to macroscopic structuring permit to be combined with the ProR approach, including [IEEE, 1997, Jackson, 2001]. In fact, some approaches, like Problem Frames [Jackson, 2001], complement the ProR approach well and are used in the case study.

To be applicable with the ProR approach, a formal method should be compatible with the predicative reasoning style employed here. The case study employs Event-B [Abrial, 2010], but also provides an example using LTL [Plagge and Leuschel, 2010]. Event-B is well-suited, because it is straight forward to specify state-based systems and to express artefacts with invariants, and it supports refinement. Further, tool support is available in the form of the Rodin platform [Abrial et al., 2010]. Rodin is well-suited for tool integration, and Chapter 4 demonstrates this by integrating Rodin with ProR, a requirements engineering platform [Jastram, 2010].

This chapter describes the theory behind the ProR approach. Chapter 5 contains a small case study, while Chapter 4 describes tool support.

3.1 Problem Statement

There are broadly two approaches to specifying systems which have complementary advantages. One approach is the use of natural language (or other informal notations). The main benefit is the ubiquity that makes it easy for all stakeholders to understand the specification, but at the expense of clarity: Keeping an informal specification correct, complete and unambiguous is labor-intensive and error prone.

On the other and, formal specifications are typically unambiguous and rigorous reasoning can be used. But they are often not comprehensible to non-experts, and often, there are some specification elements that can be expressed formally only with difficulty, or not at all.

This leads to the following problem statement:

> *Develop a practical approach for specifying systems that combines formal and informal specification methods to take advantage of their respective advantages and minimises their respective disadvantages.*

To address this problem, I identified the following properties that the approach should have. These properties are described in detail below. In Section 3.7 it is discussed whether the ProR approach succeeds.

- The approach is based on a language that the stakeholders understand (Section 3.1.1).

- The approach allows formal and informal artefacts to co-exists (Section 3.1.2).

- The approach allows the partial formalisation of the system description (Section 3.1.3).

- The approach is not tied to a specific formalism (Section 3.1.4).

- The approach supports traceability (Section 3.1.5).

- The approach supports the evolution of requirements (Section 3.1.6).

- The approach is suited for industrial use (Section 3.1.7).

Let's look at these properties one by one.

3.1.1   Stakeholder Language

A stakeholder is an individual or organisation having a right, share, claim, or interest in a system or in its possession of characteristics that meet their needs and expectations [IEEE, 2010]. Common stakeholders include customer and developer. Examples of other stakeholders include the system administrator, project manager or the quality assurance team.

Communicating effectively with all stakeholders can be challenging. The smallest common denominator is often natural language. Natural language has the big advantage of being universal and extremely rich. Many complicated requirements can be expressed concisely in natural language.

One downside of natural language is its ambiguity. There are techniques that help reduce ambiguity, but ultimately, writing clear requirements is labor intensive (see Section 2.2).

Even if individual requirements are stated clearly and without ambiguities, managing large specification is challenging. It may be hard to find an inconsistency in an 80-page specification if there is a contradiction on, say, page 5 and on page 76. Also, it takes discipline to use a well-defined vocabulary.

One way to address these concerns regarding natural language are "best practises, as can be found in [IEEE, 1997], for instance. This standard contains template outlines of requirements documents that help to identify missing content; it provides quality criteria that can be systematically applied to improve the overall quality; they recommend the creation and maintenance of glossaries; etc. Those approaches are effective and are being employed in industry. But they are labour-intensive.

However, even when natural language is used and carefully formulated, there is plenty of room for misunderstandings. Language can be ambiguous. Many domains have their own jargon. Consider the legal domain, where many common terms have a well-defined meaning. This can create problems when all stakeholders believe that they understand the requirements, but interpret them differently.

There are cases where the stakeholders have a common domain-specific language (DSL), at least for some aspects of the system under development. In that case it is preferable to use that language. DSLs could simplify the application of the ProR approach, especially if required information could be extracted from the DSL. The tool platform ProR has been successfully integrated with DSL editors [Jastram and Graf, 2011d]. This has not been investigated further in this work.

In this work, the stakeholder's language is accepted as a given, with all its potential problems. This approach attempts to compensate some of the weaknesses of the stakeholder's language, without affecting its strength.

3.1.2 Co-Existence of Formal and Informal Artefacts

Formal methods allow rigorous reasoning about a model by using mathematics. If stated formally, the use of formal methods allows to ensure that certain properties are satisfied, either by mathematical proof [Abrial et al., 2006] or model checking [Clarke, 1997]. Formal methods, however, come at a price [Berry, 1999]. This price may be worth paying for "highly safety- and security-critical systems, for which the cost of failure is death or is considered very high". But even in those cases, modelling the system as a whole formally may be prohibitively expensive. Modelling just the safety-critical subset may be sufficient to benefit from the advantages of formal methods, while keeping the costs in check.

When modelling just a subset of the system formally, it is important to understand the interdependence of formal and informal properties. As an example, consider modelling a state machine with a formalism for discrete systems modelling. Many such formalisms do not support timing properties, and it is reasonable to exclude these from the formal model. In such a case it is important not to forget to validate the timing properties as well, which has to happen outside the formal model[1].

The ProR approach takes this into consideration and allows the mixing of formal and informal artefacts.

3.1.3 Partial Formalisation

In the previous section it was stated that formal and informal artefacts can co-exist. It left open whether the formal artefacts also have an informal representation, and vice versa. This creates the following three scenarios:

The informal artefact has no formal representation. If the effort required to model an artefact formally is not justified by the result (e.g. because it is not a safety-critical property), it may be left as is. Such artefacts still have to be validated.

The formal artefact has no informal representation. If all stakeholders understand the formal representation, the informal representation can be omitted. Note that different artefacts have different stakeholders.

---

[1]Such a situation is described in the case study in Section 5, where a traffic light system for pedestrians is developed.

The group of people dealing with the specification S is probably different from those dealing with requirements R.

Formal and informal artefact coexist. Even if a concise formal representation had been found, it may not be comprehensible to all stakeholders. Therefore, the informal representation must not only be kept, but also maintained: If the formal representation is modified, the informal representation must be adjusted to reflect the new meaning.

3.1.4   Support for Multiple Formalisms

Many formalisms exist, and the ProR approach is designed to work with more than one, as long as they operate on the same model. For example, the case study in Chapter 5 expresses some system properties as invariants (Section 5.4.3) and some as LTL expressions (Section 5.5.1).

This also results in different approaches to validation. In the case study, invariants are validated with a theorem prover and LTL expressions with a model checker. Further, some properties are validated by inspection of the model (Section 5.7) or animation (Section 5.5.2).

On step further is the use of different models. For instance, in the Bosch cruise control study [Loesch et al., 2010], the project partners decided only to model the signal evaluation subsystem in Event-B, as it was well-suited for that formalism. They omitted the velocity control subsystem, which calculates an acceleration demand based on the current vehicle speed and the stored target speed. In a real project, this subsystem would be modelled in a modelling tool for dynamic systems, like Simulink.

The ProR approach supports different models in principle, by employing the same methods that are use for mixing informal and formal artefacts. However, this is not covered by the case study.

3.1.5   Traceability Support

The concept of traceability in general was introduced in Section 2.3, and the traceability between requirements and formal models in Section 2.8. The traces used by the ProR approach are described in Section 3.2.

Traceability is practised for four reasons [Gotel and Finkelstein, 1994]. These four reasons are listed below, combined with an analysis on how they relate to the problem statement, and how they are addressed in the ProR approach:

Purpose driven. Traceability exists to achieve something. It can be used to demonstrate that a requirement has been understood, or even to fulfil certain legal obligations.

Solution driven. Traceability exists to document the solution to a problem, for instance, a realisation trace from the system specification to a requirement.

Information driven. Traceability exists to connect related elements, e.g. functions and data to requirements or to each other.

Direction driven. Traceability exists to emphasise a forward or backward direction, for instance following an element through the development life cycle.

Any trace may address multiple purposes at the same time. A "realisation" trace that connects system specification elements to the requirement that they realise addresses all four reasons.

Traceability in the ProR approach has the overarching purpose of improving the quality of the system description and is therefore purpose driven. This is done by demonstrating adequacy of the specification with respect to the requirements document, while allowing to mix formal and informal artefacts. These relationships are formalised in Section 3.3 and materialise in the form of the traces introduced in Section 3.2.

Once the traceability is established, it can be used to reason about the relationship between artefacts, specifically:

Find out which artefacts have been realised or justified. This also indicates whether the realisation is formal, informal or a mixture of both.

Find out which formalisations must be validated against informal artefacts. This is necessary for artefacts with justification or equivalence traces to informal artefacts. A tool can track the traced artefacts and mark a trace as suspect ($\nrightarrow$), if the source or target of the trace have changed.

Find problems in artefacts through the model. If there is a problem with the model, it may be tracked back to a problematic artefact. If, for instance, an invariant cannot be proven by a theorem prover due to a contradiction, it may be possible to track this back to contradicting artefacts.

Manage change. Evolving the model will inevitably also change the artefacts. By analysing the suspect ($\nrightarrow$) traces, the impact of changes can be assessed, and the changed model systematically re-validated.

Find problems. Establishing a traceability can uncover problems like ambiguities or contradictions simply through the process of classifying them and/or formulating artefacts in a formal notation.

If traceability is not done right, it can be more harmful than helpful and waste valuable resources. In the ProR approach, traceability is precisely defined (what traces exist, how are they used, when are they validated). This reduces the risk of an incomplete traceability and supports the process of keeping the system description consistent.

While the focus of this work is on the traceability between artefacts, there are many more areas where traces can be used. These include project management and testing. Consequently, the tool ProR (Section 4) supports a generic form of traceability. A customisation of the tool, specifically for the ProR approach, is provided in the form of a plug-in (Section 4.7).

Requirements traceability is an established concept in industry, and most industrial-strength tools supports it [Hood et al., 2007].

3.1.6   Requirements Evolution

The initial requirements document from the stakeholders rarely stays unchanged during the specification process, as issues are identified and addressed. These include removing ambiguities, resolving contradictions, adding missing information and the like. This may result in the removal of artefacts or the addition of new ones.

In addition the ProR approach requires a certain structure, and establishing this structure will lead to more changes. This includes the identification of phenomena and the classification of artefacts into requirements and domain properties.

The evolution of the requirements document must be documented. This suggests an iterative approach that was first introduced in [Jastram et al., 2010]. Each artefact in the requirements document must be traceable to its origin in order to justify its existence.

Working iteratively has the danger that the validation of existing formalisations has to be repeated. Also, managing the traceability information manually is labor-intensive and error prone. Tool support can help. The ProR approach is supported by a tool development called ProR that is described in detail in Section 4. By marking traces as suspect and by giving users the ability to removing the suspect status by validating the trace, the effort of keeping all traces validated is minimised.

3.1.7 Industrial Applicability

A firm requirement of this work is the applicability in industry. For one, part of this work has been sponsored by the Deploy Project [EU FP7 Project, 2012], which has the explicit aim

> "to make major advances in engineering methods for dependable systems through the deployment of formal engineering methods. (...) The work of the project will be driven by the tasks of achieving and evaluating industrial take-up, initially by DEPLOY's industrial partners, of DEPLOY's methods and tools, together with the necessary further research on methods and tools."

It is my personal goal to commercialise this work upon completion of the doctorate program (Section 1.8). Therefore, industrial applicability was a core requirement.

3.2 Traceability

When discussing approaches to development, it is easy to intermix the notions of notation and method. Examples include the Event-B method and the Event-B notation, or the Problem Frames method and notation. The ProR approach differs from such approaches, in that it does not have its own notation. The ProR approach is concerned with the structure of the system description, but it does not require a specific notation. Instead, it is designed to work with any notation, including natural language.

The ProR approach allows the formalisation of a subset of artefacts, but even here, no specific notation is prescribed. Section 3.4.3 explores how a subset of artefacts could be formalised using state-based modelling and refinement.

The starting point for the ProR approach is typically an initial set of requirements from the stakeholders in natural language. A main concern of the ProR approach is the classification, structuring and extension of those initial artefacts. The foundation for such structuring is the WRSPM reference model (Section 3.2.1).

The next section introduces the reference model, followed by the extension of the mode for the ProR approach in Section 3.2.2. The result is a classification scheme for artefacts and phenomena, as well as a number of properties, of which *adequacy* is an important one. The system description should be sufficient to realise the requirements, but without being abundant.

Next, a number of relationships between artefacts is established as the foundation for the traceability. This includes *justification* of artefacts, which is used for reasoning (Section 3.2.3), *equivalence*, which is a strong form of justification (Section 3.2.4), *evolution*, which allows to follow the history of artefacts (Section 3.2.5) and *uses* to trace how phenomena are used in artefacts (3.2.6).

3.2.1   The WRSPM Reference Model

The ProR approach is based on the WRSPM [Gunter et al., 2000], which is introduced here. WRSPM is a reference model for applying formal methods to the development of user requirements and their reduction to a behavioural system specification. The modifications to WRSPM for the ProR approach are presented in Section 3.2.2.

Figure 3.1 is taken from the above paper and depicts the main artefacts of WRSPM.

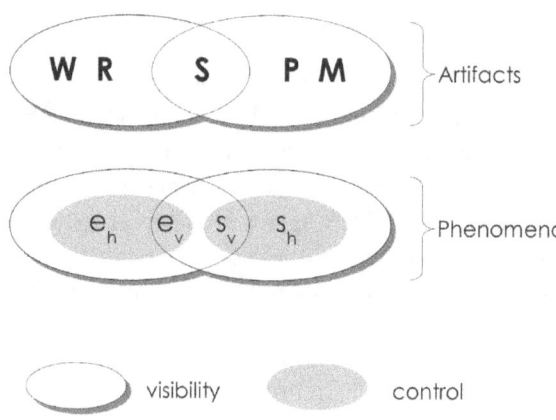

Figure 3.1: The elements of the WRSPM reference model [Gunter et al., 2000]

The artefacts are broadly classified into groups that pertain mostly to the system versus those that pertain mostly to the environment. These are:

Domain Knowledge (W) describes how the world is expected to behave.

Requirements (R) describe how the system should affect the world's behaviour.

Specifications (S) bridge the world and the system.

Program (P) provides an implementation of S.

Programming Platform (M) provides an execution environment for P.

Artefacts are written in various languages that require a problem-specific terminology. The reference model demands a clarification of the primitive terms used in the WRSPM artefacts. Terms typically designate states, events, and individuals and are referred to as phenomena.

WRSPM distinguishes phenomena by whether they are controlled by the system (belonging to set s) or the environment (belonging to set e). They are disjoint $(s \cap e = \varnothing)$, while taken together, they represent all phenomena in the system $(s \cup e = \text{"all phenomena"})$. Furthermore, they are distinguished by visibility. Environmental phenomena may be visible to the system (belonging to e_v) or hidden from it (belonging to e_h). Correspondingly, system phenomena belonging to s_v are visible to the environment, while those belonging to s_h are hidden from it. These classes of phenomena are mutually disjoint. Formally stated, this means:

$$e_h \cup e_v = e$$

$$e_h \cap e_v = \varnothing$$

$$s_h \cup s_v = s$$

$$s_h \cap s_v = \varnothing$$

The distinction between environment and system is an important one; omitting it can lead to misunderstandings during the development. In the ProR approach, the boundary is not fixed and may change depending on project characteristics. It also serves to illustrate that the boundary may be moved as development progresses. Clearly defining the boundary of the system clarifies responsibilities and interfaces between the system and the world and between subsystems. Making that distinction explicit can avoid many problems at an early stage.

W and R may only be expressed using phenomena that are visible in the environment, which is $e \cup s_v$. Likewise, P and M may only be expressed using phenomena that are visible to the system, which is $s \cup e_v$. S has to be expressed using phenomena that are visible to both the system and the environment, which is $e_v \cup s_v$.

The objective in systems development is to construct a program P, executed on a machine M which realises the requirement R, as long as the domain properties W hold. This is called *adequacy* and can be expressed as follows:

$$\forall e\ s \cdot W \wedge M \wedge P \Rightarrow R \qquad (3.1)$$

The goal of WRSPM is to decouple the implementation P from the requirement R by means of a specification S. There are practical reasons for doing this: The task of recording requirements and domain properties is typically done by a different group of people than the task of implementing the system.

A simplified but intuitive approach is to model $S \wedge W \Rightarrow R$, and that $M \wedge P \Rightarrow S$. This would decouple W and R from P and M. The first property is called *adequacy with respect to S*:

$$\forall e\ s \cdot W \wedge S \Rightarrow R \qquad (3.2)$$

This simply says that the specification constrains the world such that the requirements are realised. The trivial solution to (3.2) is obviously not interesting, meaning that no e and s exist to satisfy W and S.

Given both hidden and visible environmental (e) and system (s) phenomena, the system specification (S), under the assumption of the "surrounding" world (W), is strong enough to establish the requirements (R), which also follow from (3.2).

In addition to adequacy (3.2), consistency has to be shown as well. This goes beyond showing that a non-trivial solution exists: a property that says that any choice of values for the environment variables visible to the system is consistent with $M \wedge P$ if it is consistent with assumptions about the environment [Gunter et al., 2000]. Taking into account that P shall not be part of the property, this results in *strengthened version of relative consistency for S*:

$$\forall e_v \cdot (\exists e_h\ s \cdot W) \Rightarrow (\exists s \cdot S) \wedge (\forall s \cdot S \Rightarrow \exists e_h \cdot W) \qquad (3.3)$$

The system-side proof obligation is a similarly strengthened version of relative consistency for $M \wedge P$ with respect to S and can be found in [Gunter et al., 2000] as well.

In this work, a simpler notion of consistency is employed, based on state-based modelling, as described in Section 3.3.

3.2.2 Adoptions of WRSPM for the ProR approach

The ProR approach differs in five significant ways from WRSPM:

The ProR approach is marginally concerned P and M. P represents a program that implements the specification using the programming platform M. A good system specification S from the requirements is already very useful in practice, even without a formal extension to P. Further, including the programming platform my discourage, rather than encourage the adaption of the ProR approach: it is easier to modify existing development processes if the elements to be modified are as small as possible (while still adding value). Nevertheless, it can be useful in practice to add implementation details to the system description as shown in the case study in Section 5.7.

The ProR approach introduces design decisions D. The specification contains artefacts that represent *design decisions*. These typically cannot be justified by any of the existing WRSPM artefacts. Instead, they are introduced by the designer, who decides on a solution based on experience.

The ProR approach introduces non-functional requirements N. While WRSPM designates all requirements, in ProR approach R refers only to the functional requirements, while N represents the non-functional requirements.

The ProR approach distinguishes formal and informal artefacts. WRSPM makes no assumption about the formality of the artefacts (in fact, [Gunter et al., 2000] provides informal artefacts in an example). In the ProR approach, formal and informal artefacts may be distinguished. The ProR approach uses the superscripts F (A^F) and I (A^I) to indicate whether an artefact is formal or informal, respectively.

The ProR approach relies on refinement. For formal artefacts, the ProR approach employs refinement to iteratively add formalised artefacts to the system description.

Figure 3.2 depicts the various artefacts, and how they relate to the phenomena.

With respect to the original WRSPM model from Figure 3.1, the visibility of phenomena for W, R, S P and M did not change.

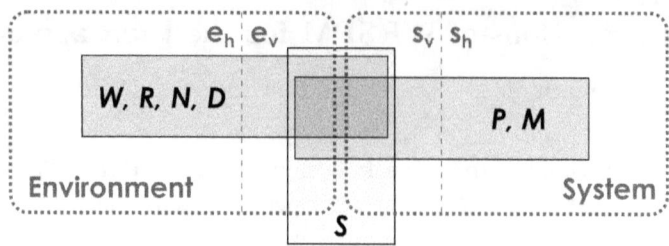

Figure 3.2: The modified WRSPM model used in the ProR approach

The design decisions D may only be expressed using phenomena that are visible in the environment, which is $e \cup s_v$. As they must justify decisions for S, they must be as expressive as S. In addition, part of the justification may be other phenomena from the environment that are not explicitly modelled (e_h). A corresponding argument can be made for the non-functional requirements.

With the extended model, *adequacy with respect to S* (3.2) takes on a new form, incorporating the design decisions:

$$\forall e, s \cdot W \wedge S \Rightarrow R \wedge D \ . \tag{3.4}$$

In addition to the requirements, now the specification constrains the world such that the design decisions are realised as well. The specification is expected to be feasible assuming a non-trivial W (meaning $\neg(\exists e, s_v \cdot W)$).

The implementation should also satisfy a condition similar to adequacy, taking design decisions D into account:

$$\forall e, s \cdot W \wedge M \wedge P \Rightarrow R \wedge D \ . \tag{3.5}$$

This can also be achieved by using the specification S instead of R and D, if adequacy (3.4) has been established:

$$\forall e, s \cdot W \wedge M \wedge P \Rightarrow S \ . \tag{3.6}$$

The latter formula (3.6) reflects the refinement condition for relations presented in [Hoare and Jifeng, 1998].

This approach distinguishes functional and non-functional requirements. The latter depend on design decisions in particular, as discussed in [Chung and do Prado Leite, 2009]. Specifically, design decisions may introduce architectural concepts or constrain the implementation. Functional requirements may add some additional non-functional, as well, they may suggest a certain technology, for instance. Therefore, R and D have to be taken into account as well:

$$\forall e, s \cdot W \wedge R \wedge S \wedge D \Rightarrow N . \tag{3.7}$$

Non-functional requirements will rarely be formal. Hence, formula (3.7) will usually consist of formal and informal artefacts with the conclusion N being informal.

The implications in the formulae (3.4) to (3.7) indicate relationships between specific artefacts. For instance, a specific specification element S_i may imply a specific requirement item R_j, therefore suggesting a realisation trace between them. The modified reference model provides the foundation for the ProR approach of requirement traceability.

3.2.3 Justification, Realisation and Satisfaction Base

An artefact B *justifies* A[2], if B justifies the *presence* of A. This can be written as $B \leftarrow A$. The underlying idea is that all artefacts that are present in the system description should be there for a reason. But more importantly, if the implication from (3.4) is read from right to left, it can be interpreted as justification relationships, specifically, $R \wedge D$ justify $W \wedge S$. In other words, the inclusion of every specification element or domain property should justified by by requirements and design decisions.

If the justification is read the the reverse direction, $A \rightarrow B$, it becomes a new meaning and can be read as A *realises* B. Again, (3.4) can be interpreted as $W \wedge S$ realise $R \wedge D$. In other words, every requirement or design decision must be realised in the form of specification elements and domain properties.

The notion of the realisation relationship corresponds even closer to the implication in (3.4), as the justification relationship corresponds to reverse implication.

For the realisation of each R and D, not all of $W \wedge S$ may be necessary. A subset SB of the artefacts $S \cup W$ is called *satisfaction base* for $R \wedge D$, if $SB \Rightarrow R \wedge D$ [Tennant, 2005, Kang and Jackson, 2010] . A small satisfaction base is advantageous for two reasons: First, a small satisfaction base provides a more precise justification than a bigger one (or $S \cup W$). Second, as a consequence of this, if the justifications have to be validated by hand, a small satisfaction base reduces the work load on the user.

---

[2]It will be clear further down why B and A are not reversed

A single smallest satisfaction base may not exist, and finding it may not be feasible, even if it exists. For practical purposes, a good estimate is sufficient.

3.2.4 Equivalence

During the formalisation of artefacts, there may be direct correspondences between informal artefacts A^I and formal ones B^F. Such artefacts are considered *equivalent*, which is denoted as $A^I \leftrightarrow B^F$.

Such a relationship is particularly useful if a formal justification for B^F has been found: In such a case, A^I is justified as well.

Unfortunately, an equivalence cannot always be found. Instead, one of the artefacts is stronger then the other. Whether this is acceptable or not depends on the artefact, with respect to the relations (3.4) to (3.7).

Domain properties W are always on the left side of the implications in those formulae. Therefore, the relation $B^F \rightarrow A^I$ reflects the implication $B^F \Rightarrow A^I$ If A^I and B^F are only related by implication following this correspondence, statements about formal world properties may not hold with respect to the corresponding informal world properties. For this to hold we need either $A^I \rightarrow B^F$, that is the formal assumption about the world are not stronger than the informal assumptions, or equivalence $A^I \leftrightarrow B^F$. Equivalence means that the informal domain properties are not stronger than needed for building the system.

For requirements, the opposite is true, as they only appear on the right side of the implications in formulae (3.4) to (3.7). If equivalence cannot be established, then the informal requirements must be realised by formal constructs B^F, as in $B^F \rightarrow R^I$.

3.2.5 Evolution

As the model *evolves* over time from A to B due to the process of modelling, changing requirements, and the like. This is written as $A \rightsquigarrow B$, for A evolves into B.

Evolution does not follow logical implication. Instead, an approximation of the change of artefacts over time can be recorded. Tracing the evolution of artefacts allows stakeholders to follow original requirements in the system description.

3.2.6   Usage of Phenomena

The traces described so far concerned the relations between artefacts, nothing so far has been said about the relation between phenomena and artefacts.

An artefact A *uses* phenomena p, written as $p \in A$. Section 3.2.2 describes, and Figure 3.2 depicts the phenomena that are permissible for use in the various artifact types. These restrictions apply to informal and formal artefacts alike.

Further, formalised artefacts may only use those phenomena that are used by the corresponding artefacts, identified by justification or equivalence. There are few means for achieving consistency between formal and informal artefacts. Usage traces are a simple but effective, and comparable to type checking or the use of alphabets in UTP [Hoare and Jifeng, 1998].

3.3   Formal Modelling and Refinement

The ProR approach allow the mixing of formal and informal artefacts. This work uses the Event-B formalism, which was introduced in Section 3.3.1. Event-B is not expressive enough to allow the formalisation of all artefacts. Therefore, formal and informal reasoning must be combined. For instance, Event-B is well-suited for state-based modelling, but not for expressing temporal and real time properties. The formalism may also influence the boundary of the system, and may even be adjusted, as more and more artefacts of the model are formalised.

Event-B supports refinement, which is used to gradually formalise more and more artefacts of the system description. This also serves as a structuring mechanism for the formal model.

Event-B uses the *proof obligations* for demonstrating consistency of the formal model, and these are used for tracing artefacts into and within the formal model (Section 3.3.2).

This ideas presented in this section have been presented in large parts in [Hallerstede et al., 2012].

3.3.1   Overview of Event-B

There are many good description of Event-B models, including [Abrial, 2010]. The following is only a brief introduction.

Event-B consists of *contexts* and *machines*. A context contains static properties in the form of *carrier sets*, *constants* and *axioms*. Axioms must not be violated, which has to be proven by discharging *proof obligations* (Section 3.3.2). A Context can be *extended* by another context, providing a simple structuring mechanism.

Machines contain the dynamic properties in the form of *variables*, *events* and *invariants*. State is represented by the variables, and events allow state transitions between states. Invariants must always hold, which also has to be proven. Machines can be *refined*, which allows the addition of more variables and invariants, as well as refinement of events. The parent of a refined machine is called *abstract machine*. The process of refinement results in additional proof obligations that ensure that neither the abstract machine's invariants are violated nor that the state space is increased by the refined events.

In addition to correctness, proof obligations for *convergence* and *deadlock freedom* can be generated.

3.3.2   Proof Obligations and Traceability

Proof Obligations are a means for identifying problems with the system description. In this section, a number of small examples demonstrate this.

Consistency Proof Obligations

Event-B generates a number of proof obligations for consistency. This includes transition proof obligations that validate that an invariant still holds after a state transition, of feasibility with respect to modelled domain properties, i.e. constants.

Consider the following domain properties that describe a list of numbers — this could be part of the domain for a sorting algorithm, for instance:

W-1	The [*list*] consists of [*N*] numbers.
W-2	The [*list*] contains the numbers [*0*]..[*N*].

This can be modelled in an Event-B context as follows:

axm1 : $card(list) = N$

thm2 : $list = 0 .. N$

The corresponding traceability is:

$$W\text{-}1 \leftrightarrow axm1$$

$$W\text{-}2 \leftrightarrow thm2$$

The proof corresponding obligations cannot be discharged, as there is a contradiction between $axm1$ and $thm2$, which can be traced to a contradition in W-1 and W-2. This contradiction can be resolved in various ways, for instance by modifying W-2:

W-2   The [*list*] contains the numbers [*0*]..[*N-1*].

With proper tool support, this change would immediately mark the trace to the $thm2$ as suspect (W-2 $\leftrightarrow$ $thm2$). After adjusting $thm2$ to $list = 0 .. N - 1$, the trace can be marked as validated.

Temporal Properties

Consistency proofs can also be used to demonstrate simple temporal properties. Specifically, Event-B generates proof obligations to validate invariants, including their initialisation. Therefore, a requirement that expresses "always R^I" can be modelled in Event-B as an invariant A^F with the traceability $R^F \to$ always R^I (or its stronger form, $R^F \leftrightarrow$ always R^I). A violation of the invariant would manifest itself in an undischarged proof obligation.

More complex temporal properties could be expressed in a different formalism. For instance, LTL [Plagge and Leuschel, 2010] can represent temporal properties of an Event-B model. While Event-B does not include a theory for creating proof obligations for LTL expressions, it is still possible to verify the properties by other means, for instance by model checking.

Refinement Proof Obligations

In Event-B, it is guaranteed that invariants from an abstract machine are preserved in their refinements (assuming that all proof obligations are discharged). Therefore, artefacts that are modelled as invariants will likewise hold in subsequent refinements. This allows the gradual enrichment of the formal system description via refinement.

Refining an event allows for *guard strengthening*, which guarantees that the concrete event cannot occur more often than the abstract event. Further, the action of the concrete event must have a corresponding effect as the action of the abstract event (*action simulation*). Last, the association of abstract and concrete event is realised by means of a *witness*. Basically, a witness associates abstract event parameters and variables to the event parameters and variables of the concrete machine. Their feasibility has to be shown as well (corresponding proof obligations are generated).

Refinement allows for *data refinement*, where an abstract state is refined in a more concrete one. This can be useful for expressing artefacts in a general way, while using refinement later to document how things work in detail. This concept is used in the case study to talk about the state of traffic lights. In the abstract machine, the traffic light state is modelled as "stop" and "go", and fundamental properties are modelled as invariants using these states. Later, data refinement is used to model these states with the actual colours of the traffic light, while a witness associates the states. The witness itself has a trace to an informal representation. Therefore, witnesses also help to make requirements traceability easier, as the proof obligations for the witness record which premises have been used.

Tracing and Correctness

It would be desirable if all traces between formal and informal artefacts were equivalences. But in practice, this is not realistic. If equivalence is not achieved, it is relevant which artefact is stronger. This in turn depends on the type of artefact, as the traceability reflects the relationship in (3.4) – (3.7). Specifically, whether the artefact in question appears on the left or the right of the implication in those formulae.

Domain properties, for instance, only appear on the left side of the implications. Therefore, the formal model must not strengthen assumptions about the domain properties, but may weaken them. Domain properties W^I must realise formal model elements A^F:

$$W^I \rightarrow A^F \tag{3.8}$$

For artefacts that are on the other side of the implication, it is the other way around. This is in particular true about requirements. Therefore, the formal model elements B^F must realise the informal requirements R^I:

$$B^F \rightarrow R^I \tag{3.9}$$

The disadvantage of realise traces, compared to equivalence traces ($\leftrightarrow$), is the fact that the system may be "better than it needs to be". The system overachieves, because the assumptions have been weakened or the requirements strengthened. Fortunately, this does not affect correctness [Apt et al., 2009].

Adequacy of the Formal Model

A central idea of this work is to show that the system description is adequate with respect to the requirements, as expressed in (3.4). For informal artefacts, this can be written as

$$\forall e, s \cdot W^I \wedge S^I \Rightarrow R^I \wedge D^I . \qquad (3.10)$$

To show adequacy of the formal model A^F, both sides of implication can be dealt with separately, using the realisation relationships, as described in Section 3.3.2 above. Not all artefacts have to be formalised, the following demonstrates the relationship only for those informal artefacts that are realised in the formal model:

$$W^I \wedge S^I \rightarrow A^F \qquad (3.11)$$

$$A^F \rightarrow R^I \wedge D^I \qquad (3.12)$$

Event-B allows the identification of formal model elements used in proof obligations. Once these are found, the traceability can be used to find the corresponding informal artefacts. If a proof obligation fails, this narrows down the set of artefacts that need to be consulted to find and correct the problem.

Tracing and Formal Refinements

As mentioned before, invariants are preserved across the refinement hierarchy. Therefore, artefacts that are realised in the form of invariants in an abstract machine will also be realised in the refining concrete machine. This allows the gradual addition of more and more artefacts through subsequent refinements. During refinement, invariants may be strengthened, however. This is fine for requirements and design decisions, but not for domain properties and specification elements (as argued in Section 3.3.2). This has to be demonstrated somehow, for instance by validating the realisation relationship against the concrete (instead the abstract) invariant.

While the same problem exists for specification elements in principle, they can, in fact, be strengthened in the formal model. But doing so would require to strengthen the corresponding informal specification element as well. This is possible, as the specification elements are being developed, and not given (in contrast to the domain properties).

Last, refinement can go beyond specifying the system, resulting in implementation detail P. Implementation artefacts may use phenomena that are invisible to the environment (s_h). While implementation is not in the scope of this work, neither the ProR approach, nor Event-B refinement makes a particular distinction. Sometimes it can be practical to include some implementation detail in the system description, as demonstrated by the case study.

Informal Proofs about Formal Models

Artefacts that are formalised as invariants or theorems are easy to verify with Event-B, as has been shown. This can also be achieved for artefacts that are realised as event guards or actions, as described further in [Hallerstede et al., 2012]. But there are cases where additional constructs have to be used, which allows formal modelling of an artefact, but not its formal validation. This is demonstrated in the case study, where a state machine is modelled formally. In that example, there are no proof obligations that verify that the correct state machine has been realised.

But the ProR approach permits and encourages informal proof as well. The informal proof can be managed in the form of annotations to the traces (something that the tool supports). In its easiest form, an informal proof is just an argument. A more sophisticated approach is the formulation of properties (e.g. temporal), as shown in Section 3.3.2. Informal reasoning can also be supported by tools that allow model checking or animation, for example.

Informal Proofs about Informal Models

Not all artefacts need to be modelled formally, which is an important feature of the ProR approach. Nevertheless, such artefacts still need justifications, and therefore justification traces (or correspondingly, realisation traces). This traceability has to be managed without the support of the formal model. It may take some more discipline to manage such artefacts, but the process does not fundamentally differ from dealing with formal artefacts — except that the proofs are informal.

There are aids for dealing with such informal artefacts. The Problem Frames approach, for instance (Section 2.5), has a way of dealing with the satisfaction of requirements (*frame concerns*).

3.4   A Process for Systems Development

The requirements engineering process can be broken down into requirements specification, system modelling, requirements validation and requirements management [Wiegers, 2003]. The primary activities are modelling and validation, with other activities playing a supporting role. Elicitation of requirements is also typically part of this process, but is outside the scope of this work, as it has little influence on modelling and validation. The high-level process for incrementally building up a system description is shown in Figure 3.3. The four activities are described in the following:

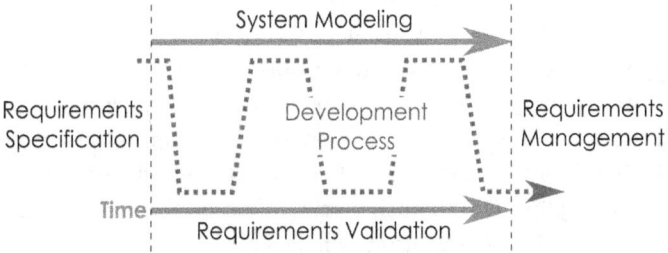

Figure 3.3: The Incremental Development Process

Requirements Specification. During this phase, artefacts and phenomena are identified and classified, as described in Section 3.2.2. Not covered by the **ProR approach** is the macroscopic structuring of the artefacts, which would take place during this phase as well. In the example in Chapter 5, the Problem Frames approach is used.

System Modelling. The objective of this phase is the creation of formal model elements representing a subset of the system description, as well as their elaboration. By using refinement, artefacts can be incorporated gradually into the formal model. This was described in Section 3.3. Not all elements need to be modelled formally, which is one distinguishing feature of the **ProR approach**. Further, any formalism can be used (see Section 3.1.4). The nature of the problem to be solved may suggest one formalism over another. It is also possible to user more than one formalism.

Modelling is an iterative process that goes hand in hand with validation, as described next.

Requirements Validation. The purpose of this activity is validating the adequacy of the specification, and the relationship between formal and informal artefacts. The validation process for informal artefacts depends on justification and realisation traces, which were created during the system modelling step. Tool support can support this step and make the validation of large system descriptions manageable. Formal model elements can be validated by mathematical proof or model checking.

Requirements Management. In practice, a specification is never "done". The ongoing work includes change management and requirement evolution. These tasks are supported by the ProR approach. Changing artefacts will result in "suspect" traces, which must subsequently be validated again. For formal artefacts, some of the validation can be performed by theorem provers and/or model checkers. The amount of formality in the system description determines how effective this is. At one end of the spectrum, all elements are modelled formally. On the other end of the spectrum is a completely informal system description, which still benefits from the ProR approach.

3.4.1   Incrementally Building the System Description

These tasks, including elicitation, analysis and negotiation, are performed in parallel. This is not a sequential process. During this process, the system description evolves: artefacts are modified, new ones are added, and traces are created. The traces employed were described in Section 3.2, and the relationship between informal and formalised artefacts in Section 3.6.

The work flow for structuring requirements is shown in Figure 3.4. The small notes indicate the kind of traces that are created in the various steps of the process.

The process starts with a set of unclassified artefacts, provided by the stakeholders. These are processed iteratively, as described in the following and depicted in Figure 3.4.

Choose Artefact

The user starts the process by selecting an arbitrary artefact. Depending on its quality, it may have to be rewritten or split, typically by checking it against a number of quality criteria (see below). If this is the case, it will result in evolution traces, and the process starts over. There is no distinction in choosing an artefact provided by a stakeholder and a new artefact, e.g. a design decision. The evolution traces should be validated by the stakeholders.

Rewriting Artefact

An artefact may be rewritten to improve its quality [Hood and Wiebel, 2005], which results in an evolution trace ($\rightsquigarrow$). The evolution trace must be validated. This is typically done by the stakeholders who confirm that the rewritten artefact still captures the original idea behind it.

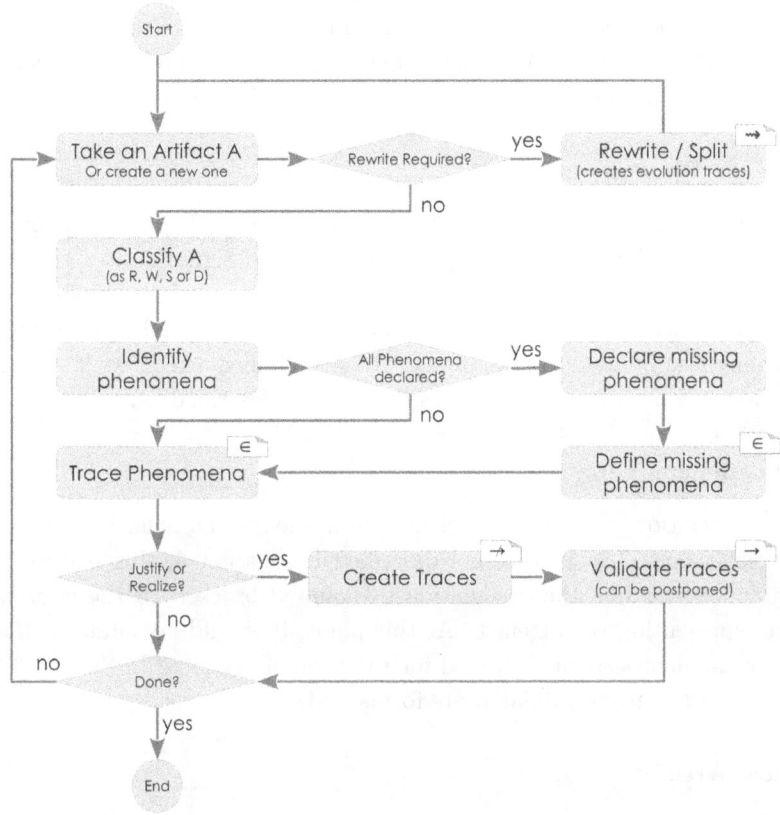

Figure 3.4: Overall work flow for building a system description, based on [Jastram et al., 2010, Hallerstede et al., 2012]. This work flow does not yet contain the process of formalisation, which is described in Section 3.4.3.

Classify Artefact

The artefact is classified as R, N, W, S or D. This in turn will determine the types of phenomena that are allowed to be used. Classifying artefacts will also fix the boundary between the system and its environment. Therefore it is important to agree on that boundary. If the boundary moves during the development, it may result in the need to reclassify artefacts.

Identify Phenomena

All phenomena in the artefact must be identified. Phenomena may or may not already declared (e.g. if the are already used by a different artefact). By identifying and classifying phenomena, a glossary is build that ensures a consistent terminology within the system description.

Declare Phenomena

Missing phenomena are declared by introducing a *designation* [Gunter et al., 2000]. This can be as simple as adding the designation to a glossary and classifying it as belonging to one of e_h, e_v or s_v. It may be described further by artefacts. Such artefacts must either be created from scratch, but may also be found in the pool of the unclassified artefacts. If such an artefact is found, it must also pass the process described here.

Trace Phenomena

The association between artefact and used phenomena must be recorded by uses traces ($\in$). In the tool described in Chapter 4, this is done by simply surrounding the designation by squared brackets in the informal text representing the artefact. At this point, it should be validated that the phenomena used are allowed for the type of artefact. In Section 3.6, the properties to be validated are formalised.

Trace Artefact

Each artefact may have justify traces ($\leftarrow$) (and correspondingly realise traces ($\rightarrow$), which are the inverse relationship). Equivalence traces ($\leftrightarrow$) are a stronger form of justification. These traces must be validated upon creation, and every time artefacts that are attached to the trace change. To postpone this validation, the traces can be marked as *suspect* ($\nrightarrow$). The traces must adhere to the constrains given by (3.4) – (3.7).

Validate Traces

All traces must be validated eventually. This is done by reviewing the relationships of all artefacts and to judge whether the construct really reflects the given justification, realisation or equivalence relationship. It is crucial that all relevant traces are included, in other words, that a correct satisfaction base has been identified. On the other hand, it is not problematic if too many traces exist, as this simply means that the satisfaction base is larger than it needs to be. If all traces have been validated, the system description is considered consistent.

Complete Iteration

This concludes the iteration cycle. At this point, the work flow either starts over, or the user considers the system description done. If declared done, at the minimum, the properties from Section 3.6 should hold.

Not mentioned in the work flow is the fact that at any point, artefacts that have already been structured may be modified again. If that happens, all justification traces connected to that artefact are marked as suspect and must be validated again. In practice, the validation of the justification traces may be postponed until the model stabilises.

3.4.2 Adequacy for Formal and Informal Requirements

System modelling as described here results in partly formalised artefacts. It is not necessarily that everything is formalised. The ProR approach proposed here allows for a mixture of formal and informal proof as a means of validation.

The ProR approach does not dictate the formalism to be used. Rather, the chosen formalism determines the amount of rigorous reasoning that is possible. The case study in this work uses the Event-B formalism (Chapter 5). The implications of choosing Event-B were described in Section 3.3.

As a consequence of frequent incremental changes, effective support for tracing artefacts is necessary: the specification changes, as it incorporates increasing detail, requirements and domain properties change as a consequence of the validation itself. The transition to requirements management is considered fluent and the same techniques of traceability are applied.

Demonstrating (3.2) now involves dealing with formal and informal elements.

In the following, R^F designates the formal requirements, W^F the formal domain properties and S^F the formal specification elements. The difference $R \setminus R^F$ of all requirements and formal requirements gives the informal requirements R^I, similarly for informal domain properties W^I and informal specification elements S^I.

Design decisions D are typically informal ($D = D^I$). From a theoretical point of view, formal design decisions are possible. The same holds true for non-functional requirements: in practice, these are also kept informal, as they are typically very hard to formalise ($N = N^I$).

For the formal elements, corresponding to (3.2) it can formally be verified that

$$\forall e\; s \cdot W^F \wedge S^F \Rightarrow R^F \;,   \qquad (3.13)$$

assuming that all W and S that are relevant with respect to R^F have been formalised, and assuming that the chosen formalism supports the rigorous verification of (3.13).

For informal elements, the corresponding relationship has to be shown:

$$\forall e\; s \cdot W \wedge S \Rightarrow R^I   \qquad (3.14)$$
$$\forall e\; s \cdot W \wedge S \Rightarrow D^I \;.   \qquad (3.15)$$

The relationship exists both for requirement R and design decisions D. Like requirements, design decisions must be realised.

Formal and informal artefacts are allowed in the antecedent of (3.14) and (3.15) but only formal elements in the antecedent of (3.13). As many critical requirements as possible should be validated formally, giving high assurance of their satisfaction. Relying on formally verified facts in informal justification will also improve their quality.

Considering that a rigorous proof of (3.14) and (3.15) cannot be provided, the relationship can only be *justified*.

Once a requirement is realised, this is documented with realisation traces. Typically, not all W and S are required for the realisation. The subset of W and S will be called *satisfaction base SB*. Each requirement and domain property has its own satisfaction base:

$$SB(R_i^I) \subseteq (W \cup S)$$
$$SB(D_i^I) \subseteq (W \cup S) \;.$$

There are many valid satisfaction bases, and $W \cup S$ is always one of them. However, the smaller SB is, the easier it is to manually validate the realisation traces. There is not always a smallest satisfaction base, nor is it feasible in practice to identify it if it exists. The concept of the satisfaction base is related to the concept of trusted bases, which has been introduced for formal models in [Kang and Jackson, 2010].

Once a requirement R_i^I or design decision D_i^I is realised, for each artefact from $SB(R_i^I)$ and $SB(D_i^I)$, realisation traces are set to document the realisation:

$$\forall A \cdot A \in SB(R_i^I) \Leftrightarrow R_i^I \to A$$
$$\forall A \cdot A \in SB(D_i^I) \Leftrightarrow D_i^I \to A$$

How the justification is performed is in the discretion of the user. The user must demonstrate that the artefacts of $SB(R_i^I)$ are sufficient to realise the requirement R_i^I, for all phenomena e and s (or more specific, only those phenomena that are used by $SB(R_i^I)$ and R_i^I). The same applies to D^I.

Under the assumption that the satisfaction base is correct (i.e. no relevant artefacts are missing) and that the justification traces are correct, adequacy of informal requirements (3.14) and design decisions (3.15) is

$$\forall e \ s \ i \cdot SB(R_i^I) \Rightarrow R_i^I$$
$$\forall e \ s \ i \cdot SB(D_i^I) \Rightarrow D_i^I \ .$$

A justification relationship becomes *suspect*, if either the source of the target of the relationship changes. Suspect justification relationships are crossed out ($\nrightarrow$). Adequacy only holds if none of the justifications are suspect.

The suspect justification link is introduced for convenience. it can also be defined as the relationship between all artefacts that are in $SB(R_i^I)$, but that do not have a justification trace to R_i^I.

3.4.3 Formalising Phenomena and Artefacts in Event-B

In order to formalise artefacts, their phenomena must be formalised first. WRSPM identifies states, events, and individuals as classes of phenomena [Gunter et al., 2000]. These correspond in Event-B to variables, events and constants. The formalisation process consists of the creation of those Event-B elements and their classification as e_h, e_v, s_h or s_v.

Variables and constants must be typed by invariants or axioms, respectively. Those are artefacts and must be classified as W, R, S or D. Creating constants may require the creation of sets as well. Such sets should receive a meaningful designation. Creating variables requires the creation of events for changing their state.

Once all phenomena for a given artefact are formalised, the artefact itself can be formalised as invariants or events. As such, traces between the artefact and the invariants and events must be created, and the Event-B elements must be classified as W, R, S or D. Central to the formalisation in the form of events is the before-after predicate, as mentioned in Section 3.3, as it allow the Event-B formalisation to fit into the the the shape of adequacy (3.2).

The trace may either be an equivalence trace or justification traces. If it is an equivalence trace, then both elements must be of the same artefact type (e.g. $R^I \leftrightarrow R^F$, $W^I \leftrightarrow W^F$ and so on). As detailed in Section 3.6, justification traces exist between $R \cup D$ and $W \cup S$.

The formal model must not strengthen domain properties. If an equivalence cannot be established, then the formal artefacts A^F must justify the informal domain properties W^I:

$$A^F \leftarrow W^I . \tag{3.16}$$

For requirements, it is the other way around: If equivalence cannot be established, then the informal requirements R^I must justify the the formal artefacts B^F, resulting in

$$R^I \leftarrow B^F . \tag{3.17}$$

In other words, the formal model must not weaken the requirements.

Equivalences are preferable, as they ensure that not more then necessary is implemented. The risk with (3.16) and (3.17) is, that either the assumptions were weakened or the requirements strengthened. Note that this would not result in a faulty specification, but rather to a system that is better than it needs to be.

The creation of formal artefacts may require further modifications to the model to keep it sound. Invariants, for instance, may require the introduction of guards to prevent them from being violated (a theorem prover or model checker can find such violations).

The soundness of invariants can be proven. If all relevant realisation traces for an artefact A exist and A is realised by invariants only, and if the realisation traces have been validated, then if the model has been proven correct, A is validated as well.

The soundness of events cannot be proven, only the soundness with respect to the before-after predicates. This is not always sufficient. For instance, later refinements may strengthen the guard, thereby breaking the realisation relationship. Thus, if an event has realisation relationship to an artefact A and is modified in a refinement, the refined event must also be added to the realisation relationship.

Rather than creating a high number of traces from an artefact A to the corresponding invariants and/or events, the trace can be linked to a refinement that encapsulates all formal elements that realise A. This approach tends to create too many traces, but this is not problematic in terms of justifying artefacts, as has been described in Section 3.4.2. However, if the refinement contains model elements beyond invariants or before-after predicates, then the tool-supported validation may be limited.

3.4.4 Other Formalisms

The ProR approach is not limited to formalisation in Event-B. The case study in Chapter 5 demonstrates the formalisation of one artefact in LTL, for instance. In general, the formalism should be suited to the problem at hand (and particularly for those artefacts that are targeted for formalisation).

The ProR approach can also be combined with semi-formal notations, like UML [Fowler and Scott, 2000], which are widely used in industry. Such notations allow the modelling of phenomena. However, UML has only a limited ability to express properties of the model. These are limited to cardinality of relationships or attribute types when using class diagrams, or state transitions when using state diagrams. But this in itself can be useful, especially if tool integration makes it easy to validate the uses ($\ni$) relationships.

With the availability of UML-B [Snook and Butler, 2006], UML modelling could be combined with Event-B modelling. This approach has not be explored further in this work.

3.5 Macroscopic Structure

Both WRSPM, and the modified version used in this work, classify artefacts and their relationship to each other. But the artefacts are merely referred to as sets, nothing is said about their presentation to the user. In practice however, this is highly relevant: A list of requirements is much easier to understand if some thought went into creating a meaningful order, as argued by [Kovitz, 1998], for instance. Further, additional structure in the form of sections, headlines, information text, etc. improves readability and scalability.

A lot of practical advice with respect to macroscopic is available [Rupp, 2007, Pohl, 2007, Hood and Wiebel, 2005]. Some of this advice is manifested in the form of standards (e.g. [IEEE, 1997]) or part of process framework templates (e.g. [Kruchten, 2004]). IEEE 830, for instance, is a document-centred approach that provides standard document outlines for different types of system descriptions, combined with some quality criteria and checklists for completeness. It does say little with respect to the artefacts themselves. But this in turn makes it easy to combine it with the ProR approach, which is primarily concerned with the artefacts and their structure and relationships.

There are methods that are concerned both with the macroscopic structure, and the internal structure of the artefacts. Those methods tend to be compatible with WRSPM. This is no accident, as WRSPM is meant as a reference model that is specifically meant for discussion of common aspects of different methods. Examples of such methods include KAOS [Darimont et al., 1997] or the Problem Frames approach [Jackson, 2001] (see also Chapter 2.5). As the Problem Frames approach is well suited to be combined with the ProR approach, and as the case study in Chapter 5 employs the Problem Frames approach for this purpose.

3.5.1   Problem Frames

The Problem Frames approach [Jackson, 2001] is a concrete approach to software requirements analysis. Problem Frames can be interpreted in the context of the WRSPM reference model.

Central to the Problem Frames approach is the idea that user requirements are located in the real world, which should behave in a certain way (once the requirements are realised). Therefore, it is crucial to first describe the world and its behaviour. The description of the world corresponds to W in WRSPM and the environmental phenomena e.

The Problem Frames approach uses *problem diagrams* to visualise this. An example is shown in Figure 3.5 (which uses dotted borders to visualise the interpretation in terms of WRSPM-terminology). The domains are visualised as boxes (Pedestrians, Cars, Street, etc.) that have certain shared phenomena, that are expressed as lines between domains. The behaviour of the world can be further specified by an optional description, which is visualised as an oval.

In the Problem Frames approach, user requirements are expressed by their effect on the world. This corresponds to R in WRSPM and is visualised as dotted ovals in problem diagrams. Requirements never relate to the actual system to be designed, but only to the domains. requirements

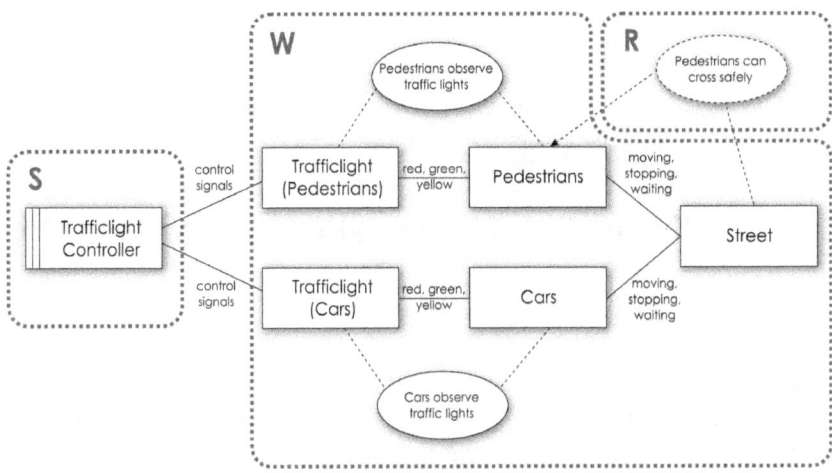

Figure 3.5: Example of a Problem Frames diagram, including mapping to WRSPM [Jastram et al., 2011]

describe or constrain the behaviour of the domains.

In order to realise the requirements, a system is designed. This is called machine domain and visualised by a box with two vertical bars. A problem diagram contains exactly one machine domain. The machine domain communicates with the world through phenomena, represented as connecting lines. In terms of WRSPM, these phenomena must be in e_v or s_v — environmental phenomena visible to the system, or system phenomena visible to the environment. Correspondingly, phenomena that are not connected to the machine domain belong to e_h: environmental phenomena hidden from the system.

The description of the machine domain corresponds to S in WRSPM: It describes how the machine must behave to realise the requirements, assuming that the world behaves as expected.

The Problem Frames approach recognises the importance of decomposition. A hierarchical decomposition of the problem is considered bad, as it takes no explicit account of the problem to be decomposed [Jackson, 2001]. Instead, it proposes the decomposition into sub-problems of recognisable and familiar classes, which are called *Problem Frames*. This approach leads to known problems, allowing to take advantage of prior knowledge. For this to work, sub-problems must be complete, so that superimposing the sub-problems won't invalidate them. At the same times, this leads to a parallel (rather than hierarchical) structure. Concurrency and possible interactions of sub-problems have to be taken into account.

Ultimately this leads to a collection of interacting sub-problems, each of which is smaller and simpler than the original, with clear and understandable interactions [Hall et al., 2002].

3.6 A Formal Meta-Model of the System Description

Both, the modified WRSPM model (Section 3.2.2) and the relationships described in Section 3.2 can be modelled formally. In this section, such a model is established. This also leads to some properties of correctly modelled system descriptions. These properties could be validated.

A system specification that is structured according to the ProR approach should, at a minimum, have the properties described here. While this does not guarantee a complete system description, their absence indicates incompleteness.

3.6.1 Formalising the Modified WRSPM

The system description consists of artefacts and phenomena. The four artefacts are R, W, S and D, while the phenomena are e_h, e_v, s_h and s_v (see Section 3.2.1).

$$partition(Artefacts, R, N, W, S, D)$$
$$partition(Phenomena, e_h, e_v, s_v, s_h)$$
$$partition(e, e_h, e_v)$$
$$partition(s, s_v, s_h)$$

Artefacts are expressed in terms of phenomena, they *use* them. Each artefact uses at least one phenomenon:

$$uses \ \in Artefacts \leftrightarrow Phenomena$$

But there are further constraints: domain properties may only use environmental phenomena, requirements may use visible system phenomena in addition, and specification elements may not use hidden environmental or system phenomena. Design decisions were introduced in see Section 3.2.2 and may use all except hidden system phenomena. Implementation detail can use hidden system phenomena, but no hidden environmental phenomena:

$$uses[W] \subseteq e \tag{3.18}$$

$$uses[R] \subseteq e \cup s_v \tag{3.19}$$

$$uses[N] \subseteq e \cup s_v \tag{3.20}$$

$$uses[S] \subseteq e_v \cup s_v \tag{3.21}$$

$$uses[D] \subseteq e \cup s_v \tag{3.22}$$

$$uses[P] \subseteq e_v \cup s \tag{3.23}$$

Likewise, each phenomenon is used by by at least one artefact. Hidden environmental phenomena are used by domain properties, requirements or domain decisions. Visible environmental and system phenomena can be used by any artifact, and hidden system phenomena only by implementation detail:

$$used\_by \in Phenomena \leftrightarrow Artefacts \tag{3.24}$$

$$used\_by[e_h] \subseteq W \cup R \cup N \tag{3.25}$$

$$used\_by[e_v] \subseteq W \cup S \cup D \cup R \cup N \tag{3.26}$$

$$used\_by[s_v] \subseteq W \cup S \cup D \cup R \cup N \cup P \tag{3.27}$$

$$used\_by[s_h] \subseteq P \tag{3.28}$$

The last four formulas (3.25) to (3.28) are actually theorems that can be drived from the former relationships.

This is the formal description of the modified WRSPM used by the ProR approach. If formal model is built using this structure, then adequacy with respect to S (3.2) could already be expressed and, if desired relative consistency for S (3.3).

3.6.2 Formalising Justifications and Realisations

Justifications and realisations are a new concept in the ProR approach that does not exist in WRSPM. To be realised, a requirement must justify at least one specification element.

Likewise, each specification element must have a reason to exist: the reason may be the realisation of a requirement. However, it may also exist to add design information to the specification. The realises relationship is the inverse of the justifies relationship.

$$justifies \in (R \cup D) \leftrightarrow (S \cup W) \qquad (3.29)$$
$$realises \in (S \cup W) \leftrightarrow (R \cup D) \qquad (3.30)$$
$$realises = justifies^{-1} \qquad (3.31)$$

3.6.3 Formalising the Distinction between Formal and Non-Formal Artefacts

The artefacts R, W and S may be formal or non-formal:

$$partition(R, R^F, R^I)$$
$$partition(W, W^F, W^I)$$
$$partition(S, S^F, S^I)$$

The design decisions D and non-functional requirements N are assumed to be non-formal only. While technically not necessary, this limitation reflects the nature of the artefacts. Likewise, implementation detail P is assumed to be formal only.

To allow formalisation of R, W and S, a number of additional properties must hold. For instance, the artefacts that are used by a formalised artefact must have been formalised. This can be expressed as follows for formalised requirements R:

$$\forall r \cdot r \in R^F \Rightarrow (\forall p \cdot p \in uses[R^F]$$
$$\Rightarrow (used\_by[\{p\}] \cap (W^F \cup S^F \cup D^F)) \neq \varnothing \quad (3.32)$$

Corresponding properties exist for W^F and S^F.

If a tool chain is used that supports syntactical validation, like Rodin (Section 4.3.1), then (3.32) holds if no syntax errors are reported by the tool.

3.7   Discussion

This chapter described the ProR approach in detail, which consists of a theory of traceability, and a process to put it to work. It allows the incremental building of a system description consisting of informal and formal artefacts.

The ProR approach structures the artefacts by classifying them and by identifying the phenomena used by the artefacts. This means that the global structure of the artefacts (e.g. the arrangement of artefacts in a document) is not affected. This in turn makes integration with existing processes relatively easy, as they typically say little about the internal structure of the artefacts. This has been discussed in Section 3.5.

The formalisation presented allows the use of theorem provers, model checkers or animators to support the validation of the system description. Nevertheless, the traceability eventually connects informal artefacts to other informal artefacts, as expressed in (3.10). This begs the question where the added value of building a formal model comes from. The value comes from the placing of the formal model between informal requirements and informal domain properties, with respect to the traceability. Rather than identifying the formal model elements with requirements or domain properties, they are merely traced to. This keeps them accessible to stakeholders, while using the formal model to make the relationship (3.10) explicit. The question of the value of formalisation is not limited to this approach, but to formal modelling in general.

Section 3.1 established a number of criteria for success. In the following, these criteria are evaluated:

The approach is based on a language that the stakeholders understand. The approach works with artefacts in natural language. The classification as W, R, N, S, D or P does not inhibit their understandability. While the phenomena used must be identified somehow, this could be done with a non-intrusive measure (e.g. by underlining phenomena in the requirements text). Even if something other than natural language is used, understandability should be achievable, as the ProR approach does not change the representation of artefacts, but merely labels them.

The approach allows formal and informal artefacts to co-exists. By creating realisation traces and satisfaction bases, traceability can be achieved, independent of whether the traced artefact of formal or informal. The traceability allows the systematic validation of artefacts. Formal and informal artefacts share the same phenomena, and therefore the same vocabulary (designations). If formal and informal artefacts are synonyms, an equivalence trace can ensure that the relationship is re-validated when necessary.

The approach allows the partial formalisation of the system description. Corresponding to the previous item, the ProR approach supports partial formalisation.

The approach is not tied to a specific formalism. As long as the formulae from Section 3.2 can be expressed with the chosen formalism, the formalism is adequate to be used with the ProR approach.

The approach supports traceability. Traceability is clearly defined, its use is described, and its purpose clear. In addition, due to the structuring of the system description, maintenance of the traces is manageable, and tool support is available.

The approach is suited for industrial use. Unfortunately, applicability for industrial use could not be verified, as the case study in Chapter 5 is not representative for industrial applications. The jury is still out on this one.

Tool support is central in making an approach, like the ProR approach, practical. In the next chapter the ProR tool is introduced, which will be employed in the case study as well.

Chapter 4

ProR Requirements Platform

A major contribution of this work is the development of a platform for managing natural language requirements, called ProR. An important aspect of the approach described in this thesis is the ability to scale beyond toy examples. This is difficult to achieve without tool support.

ProR is an Eclipse-based application that has a strong focus on supporting extensibility and integration. Further, the tool is based on the emerging Requirements Interchange Format (Section 4.2), which gives us interoperability with industry-strength requirements tools.

I developed ProR to survive beyond this dissertation and beyond the life of the Deploy project. I therefore tried to involve other parties and engaged in community building. I succeeded in attracting plenty of interest and, even more important, contributors who added code to the system. In June 2011, ProR was submitted as part of the Requirements Modeling Framework (RMF) to the Eclipse Foundation. in August 2011, RMF became an official Eclipse Project with ProR being the official name of the GUI. Community building is described in detail in Section 4.1.3.

4.1 A History of ProR

Before getting into the technical details of ProR, I will provide a brief overview of the various development stages that ProR went through, which were:

Initial Development (April – June 2010) In this period, I was designing and building a minimal working system. At the same time, I started to establish a community.

Collaboration with Verde (July 2010 – February 2011) I joined forces with the Verde research project, which provided the ReqIF core, while I provided the user interface. In both projects, there was a strong focus on getting ProR sufficiently complete to be used in traceability research.

Eclipse Foundation Submission (March 2011 – November 2011) The collaboration with Verde resulted in the successful submission of ProR and the Verde ReqIF core to the Eclipse Foundation, resulting in the creation of the Eclipse Requirements Modeling Framework (RMF).

Development as part of Eclipse (December 2011 – today) After its creation, RMF became an *incubation* project at the eclipse foundation. The project will stay in this state until full compliance with the Eclipse project requirements is achieved. There are currently public integration builds every two months.

4.1.1   Initial Development

Development on ProR started in April 2010 by myself. Within three months, I had a rudimentary requirements tool that implemented a subset of the ReqIF data model. The tool used the native EMF persistence mechanism, rather than the correct ReqIF format.

Even though EMF supports XML-based persistence and ReqIF is an XML-based format, it is not easy to tweak EMF to write correct ReqIF. However, EMF is modular and allows fairly easily to swap one persistence engine out for another. Therefore, I considered it a low risk to start development without the ability to read and write correct ReqIF.

This turned out to be a good decision. When I joined forces with the Verde-project, I did exactly that: I swapped my persistence engine for the one developed by Verde. It turned out to be a non-trivial task to create such an engine (see Section 4.4.8), but swapping one engine for another was easy.

Parallel to development, I engaged in systematic community building, which is described in Section 4.1.3.

At this point, the code repository was not published yet. It was not clear yet which license would be used.

4.1.2 Collaboration with Verde

The project got a big boost when in July 2010 ProR development got support from itemis[1], an IT services company specialising in Eclipse-based software. itemis was also developing a RIF/ReqIF tool, also for a research project (ITEA Verde), and also as an open source effort. They already had a working RIF 1.2 back end. They had a GUI as well, but it was essentially the unmodified default EMF editor, which is essentially unusable for all practical purposes.

Switching out the back end of ProR was done swiftly. Thereby ProR became a RIF 1.2 tool, instead of a ReqIF tool until that point. The EMF-based file format was still available, as the Verde back end was not able to handle all data structures correctly. For instance, SpecRelations were not restored correctly. Supporting more than one version was easy, however, and users could switch between versions by simply using the "Save as..." function of the tool. The Verde core and the technical challenges are described in Section 4.4.8.

During this time, the two projects merged repositories and grew closer together. There were several joint work sessions, collaborative papers [Jastram and Graf, 2011b, Jastram and Graf, 2011d, Jastram and Graf, 2011c], and an industry presentation [Jastram and Graf, 2011a], significantly increasing visibility or our work.

Shortly before the submission of our work to the Eclipse Foundation, the decision was made to switch from RIF 1.2 to ReqIF 1.0.1. This was shortly after a meeting with representatives of requirements tool vendors. At that meeting, it was clear that the current RIF 1.2 implementations have a poor interoperability. There was little interest in fixing this, but instead most tool vendors committed to supporting ReqIF 1.0.1 within 12 months or less. As our tool chain was several months away from being a feature-complete tool, it felt that focusing on ReqIF would be the right strategic decision.

4.1.3 Community Building

It is a tragedy that lots of high-quality research and many promising software projects never gain traction. This is true for academic work, as well as hobby projects. This is the more tragic, as there are now many successful open source projects and ideas that started as grassroots movements, and where their success has been documented and is well understood. As one goal of the work presented here is commercialisation,

---

[1] http://www.itemis.com

it is imperative that the work does not die a slow death after completion of this dissertation.

Figure 4.1: The pror.org website, before migration to the Eclipse Foundation

A central pillar of community building was the establishment the pror.org website (Figure 4.1), containing:

End User Documentation – A Tutorial, description of the user interface, features, limitations, etc.

Screencast – A screencast that shows installation and usage of ProR.

Developer Documentation – Architecture, important technical details, instructions on how to extend the tool, etc.

Blog – I made sure to post relevant articles 1-2 times a month.

Newsletter – Most blog articles were also mailed to newsletter subscribers, therefore allowing me to reach out to interested parties when new were available.

Bug Tracker – To allow users and developers to report bugs.

After the migration of ProR to the Eclipse Foundation, as described below, the content shown in Figure 4.1 was migrated to the foundation and the URL redirected.

4.1.4 Eclipse Foundation Submission

In March 2011 the project was ready to be taken to the next level and to apply to become an official Eclipse Foundation project. As there was much more than just ProR, it was decided to call the new project Requirements Modeling Framework (RMF), consisting of two major components: The RIF/ReqIF core and the ProR GUI. In June 2011, the first version of the proposal was submitted. The proposal can be found in Appendix A.

In November 2011, RMF finally became an official Eclipse Project, albeit still in the incubator status. During the migration process, it underwent a rigorous intellectual property review. The trademark right to ProR and the pror.org domain were also given to the Eclipse Foundation.

At this time of this writing, RMF is an active project. Public integration builds are released every two months.

4.2 The Development of the Requirements Interchange Format

RIF/ReqIF [OMG, 2011] is an emerging standard for requirements exchange, driven by the German automotive industry. It consists of a data model and an XML-based format for persistence.

This section provides some background regarding the standard. The format itself is described in Section 4.4.1.

4.2.1   History of the RIF/ReqIF Standard

RIF was created in 2004 by the "Herstellerinitiative Software" (HIS[2]), a body of the German automotive industry that oversees vendor-independent collaboration. Within a few years, it evolved to the version 1.2. The format gained some traction in the industry, and a number of commercial tools claims support it. The reality looked slightly different, unfortunately. Many RIF 1.2 implementation were not complete, supported only a subset of RIF features, and often worked only when used in exchange with the same tool, and sometimes not even then. Still, the marketing departments of tool vendors happily wrote "RIF support" in their promotional materials, and there was a continuous interest in the topic.

In order to quantify the interest in RIF/ReqIF, I looked at two metrics. The first was the number of talks at ReConf[3], a commercial requirements engineering trade show (Table 4.1). ReConf has a high attendance rate amongst representatives of the automotive industry (both OEMs and suppliers), and the organiser, HOOD Group[4], was involved in the development of the RIF/ReqIF standard. Further, with over 300 attendees, ReConf is one of the biggest commercial requirements events in Europe.

Year	Number of talks (ReConf)
2006	0
2007	1
2008	1
2009	1
2010	0
2011	3
2012	3

Table 4.1: Number of RIF/ReqIF talks at ReConf

The second metric is the number of scientific publications on Google Scholar (Table 4.2). By searching for "Requirements Interchange Format", the name change from RIF to ReqIF, as well as the ambiguity with other acronyms was not an issue.

---

[2]http://www.automotive-his.de/
[3]http://www.reconf.de
[4]http://hood-group.com
[5]Estimate. Count as of May 2012 is 6.

Year	Number of papers (Google Scholar)
2006	1
2007	5
2008	10
2009	8
2010	11
2011	13
2012	15 (estimate)

Table 4.2: Hits for "Requirements Interchange Format" on Google Scholar

Both metrics show some interest right after the creation of the standard, which kind of ebbs down around 2009/2010, just to pick up again.

In 2010, the Object Management Group (OMG[6]) took over the standardisation process and released ReqIF 1.0.1 in April 2011. The name was changed from RIF to ReqIF to prevent confusion with the Rule Interchange Format, another OMG standard, while the version number was reset.

ProR was initially based on RIF 1.2. An attempt to refactor ProR to support multiple RIF versions was abandoned again. Now ProR supports ReqIF 1.0.1 (see Section 4.1.2).

4.2.2   The Future of ReqIF

There are a number of activities that indicate that ReqIF is here to stay. Whether it will become a niche standard in the German automotive industry or a standard that will leave a huge imprint on system development is to be seen.

Currently, users and vendors of major requirements tools collaborate to ensure interoperability of their respective ReqIF implementations. This collaboration is coordinated by ProSTEP, a non for profit organisation[7]. All vendors of tools pledge to support ReqIF in their tools by the end of 2012. This includes IBM's tool Rational DOORS, which is widely used in the German automotive industry.

Strong support for ReqIF originates from the German automotive industry, which relies heavily on external suppliers. Their concern to be

---

[6]http://www.omg.org

[7]Information regarding the ReqIF implementer forum can be found at http://www.prostep.org/en/projects/internationalization-of-the-requirements-interchange-format-intrif.html

cornered by proprietary solutions prompted the development of ReqIF in the first place. Several OEMs are members of the ProeSTEP implementer forum and actively encourage the development of a ReqIF ecosystem.

Last, Eclipse provides a number of tools for system development, but to date tool support for requirements engineering has been week. The Eclipse Foundation provided positive feedback to the RMF project, as it promises to fill this niche. Committers of the RMF project, myself included, are actively participating in two Eclipse Working Groups. The Eclipse Working Group Automotive is an initiative for developing Automotive Software Development Tools[8], and RMF may become part of their tool platform. The Polarsys working group[9] was initiated by the aviation industry and also shows strong interest in RMF. In particular, the Topcased tool for system development is already in use in the aviation industry, and an integration of RMF and Topcased is being considered [Jastram and Graf, 2011b].

4.3   Goals for ProR

It would be hypocritical to preach about proper requirements engineering, but then not to practice it.

Elicitation resulted in a number of high-level goals for ProR. These include the following:

Rodin Integration As our work was in part sponsored by the Deploy project, one requirement was support for the Rodin platform. Rodin is an Eclipse-based application, suggesting to provide requirements management in the form of an Eclipse plug-in. Other architectures would be possible (e.g. a web application operating directly on the Rodin database), but not as straight forward to implement.

RIF/ReqIF support Using RIF instead of a proprietary data model was one of the core requirements. This gives us both interoperability with industrial-strength tools, and increases visibility.

Use outside Rodin possible ProR was not supposed to be a Rodin-specific tool, as there is a much wider audience for our tool.

Seamless integration with other tools possible Integration of other tools with ProR is encouraged – foremost with Rodin. This is another reason why Eclipse was an attractive foundation for ProR.

---

[8]http://wiki.eclipse.org/Auto_IWG
[9]http://wiki.eclipse.org/Polarsys

The plug-in and extension point mechanism of Eclipse is well-suited for realising tool integrations. There is also interest from the Topcased-community [Jastram and Graf, 2011b].

Scalability In order to be interesting to industrial users, ProR has to support large specifications. In the automotive industry, for instance, specifications with tens of thousands of requirements, each with dozens of attributes, are not unusual.

Longevity ProR was designed specifically to become a product that would survive beyond this dissertation, requiring the establishment of a community.

Industry-Strength ProR is targeted beyond the academic market. In fact, the author started a business to commercialise ProR.

Each of these goals was spanning a number of requirements. The requirements were kept as light as possible and as complete as necessary. The following outlines the requirements that follow from the goals:

4.3.1 Rodin Integration

The work described here is sponsored in part by the EU-project Deploy. Developing support for requirements management was one of the project's goals. Therefore, ProR had to support the goals of that project. There was a keen awareness that traceability between requirements and formal methods was an important research area. This results in the following requirements:

1. ProR is an Eclipse application.

2. ProR uses an EMF data model, amongst other things to allow integration with Rodin by using the Rodin EMF plug-in.

3. ProR will allow extensions that incorporate Event-B model elements into the requirements.

4. ProR will allow extensions that customise the appearance of attributes (e.g. to properly render Event-B elements).

5. ProR will allow manual incorporation of Event-B model elements via drag and drop.

6. ProR will allow plug-ins to listen to changes in the RIF/ReqIF model.

4.3.2   Seamless integration with other tools possible

Even though an integration with Rodin was necessary from the very beginning, ProR was not supposed to have any dependencies to Rodin, for multiple reasons: For one, Rodin contains many features that most users of ProR do not care about. ProR does not need to include those. It may also have created licensing issues, as is discussed in Section 4.1.4. Further, with too many dependencies, ProR could become a maintenance nightmare, where I would have to accept many decisions by the Rodin team (e.g. regarding Java or Eclipse version).

Even though ProR was explicitly separated from Rodin, an integration was imperative to support the goals of the Deploy project. The architecture therefore had to allow a tight integration of the tools with additional plug-ins. This in turn lead to an extensible architecture from the very beginning. The supporting requirement is:

1. All requirements from *Rodin integration* will be available in a generic fashion, thereby allowing integration with any EMF-based tool.

4.3.3   RIF/ReqIF support

Supporting RIF/ReqIF was one of the key decisions for commercial exploitation and community building. In April 2011, ReqIF became an official OMG standard and created a lot of buzz in the requirements engineering industry, especially the car industry. At that point, no open source implementation of ReqIF existed yet. My goal was to attract attention by being the first serious ReqIF implementation in the open source.

While ReqIF was designed to be a file format, it has the potential of being much more than that. Therefore, I used the ReqIF data model as the foundation for a tool (see Section4.4.1). This meant that ProR could read and write ReqIF without having to convert from or to a native data format. This gives it an advantage over existing tools, as it can handle all ReqIF features "out of the box".

I attended a meeting in July 2011 of the major requirements tool manufacturers that was organised by the vendor-neutral ProSTEP non-profit organisation. The goal of the meeting to ensure interoperability of the tools regarding the ReqIF standard. At that point, ProR was already recognised as a competitive tool, otherwise I would not have been invited to the meeting. It was a clear sign that this strategy paid off.

Our requirements supporting this goal are:

1. ProR supports at least one RIF/ReqIF standard in the sense that all model elements can be read and written.

2. Eventually, all model elements are shown and editable in the GUI.

4.3.4 Use outside Rodin possible

This goal complements Section 4.3.1: While the Rodin integration must be taken into consideration, ProR must not be tied to it.

1. ProR will have no dependencies to Rodin (all dependencies for the Rodin integration will reside in the integration plug-in).

2. ProR will be made available both as an update site as well as a stand-alone application.

4.3.5 Longevity and Public Support

A non-technical goal was to get public support and endorsements. This can work on two levels: Endorsement of the software's code and endorsement of the software's functionality. I succeeded on the first and are still working on the second.

OpenSource projects do not need to belong to an organisation. There are countless open source projects on Sourceforge, GitHub, or privately hosted repositories. But even privately hosted projects typically have an association by using a standard open source license. Using a standard license gives users already some confidence regarding the legal requirements on using the code.

Some credibility also comes from using a public repository like Source-Forge or GitHub. These repositories typically require projects to use of a public open source license, and due to the public nature of these sites, it is not possible to retroactively change the license.

But most credibility comes from becoming an official project of a non-profit organisation that takes over the patronage of the project. Well-known organisations include the Apache Foundation or the Eclipse Foundation. These organisations make sure that the origin of every line of code is known; that certain quality standards are applied; that patent issues are handled correctly; and so forth. They also provide legal advice and defend the software, if necessary, against lawsuits. In addition, some of these organisations have a good reputation and increase the visibility of projects significantly.

I got endorsement of the code by making ProR part of a newly formed Eclipse Foundation project. As of this writing, The Requirements

Modeling Framework (RMF) is in the incubator stage of the Eclipse
Foundation. Requirements supporting this goal are:

1. ProR is reasonable well documented for developers.

2. ProR has documentation and a tutorial for plug-in developers.

3. Provision and maintenance of an infrastructure for development (bug
   tracker, wiki for documentation, developer mailing list, user forum).

4. Become part of a non-profit body like the Eclipse Foundation, that
   provides visibility, guidance and infrastructure.

4.3.6   Industrial Strength

As one of the goals is the commercialisation of ProR, industrial acceptance
has to be ensured, which means at least the following:

1. The scalability requirements are a prerequisite.

2. RIF/ReqIF support is complete (see RIF/ReqIF support, Section 4.3.3).

3. The user interfaces passes acceptance by industrial representatives.

4. ProR is documented for the end users.

4.3.7   Scalability

Scalability is one central requirement for industrial acceptance. During my
work in requirements engineering, I got exposed to real-world requirements
and specifications. For instance, I encountered a supplier who got regularly
specifications from OEMs for instrument clusters — the integrated element
that contains speedometer, tachometer, and various other indicators. The
typical size was 30,000 – 40,000 requirements, where a requirement could
be a piece of text or an image.

The following requirements capture this in a qualitative manner:

1. ProR will be able to manage tens of thousands requirements with
   dozens of attributes in a single RIF/ReqIF file.

2. ProR will be able manage RIF/ReqIF files in the double-digit
   megabyte size.

4.4 Technologies

This Section describes the technologies employed in building ProR. This includes the RIF/ReqIF standard itself and an introduction of the Eclipse platform, which is the application framework used for building ProR. Eclipse programs are always written in Java. ProR uses the Eclipse Modeling Framework (EMF) for building the data model and the user interface. Eclipse applications have their on GUI toolkit called SWT/JFace. This had to be augmented with a third-party component called AgileGrid to make all desired features possible.

4.4.1 The Content and Structure of a ReqIF Model

In general terms, a ReqIF model contains attributed requirements that are connected with attributed links. The requirements can be arbitrarily grouped into document-like constructs. I'll first point out a few key model features and then provide more specifics from the ReqIF specification [OMG, 2011].

A *SpecObject* represents a requirement. A SpecObject has a number of *AttributeValues*, which hold the actual content of the SpecObject. SpecObjects are organised in *Specifications*, which are hierarchical structures holding *SpecHierarchy* elements. Each SpecHierarchy refers to exactly one SpecObject. This way, the same SpecObject can be referenced from various SpecHierarchies.

ReqIF contains a sophisticated data model for *Datatypes*, support for permission management, facilities for grouping data and hooks for tool extensions.

The ReqIF Top Level Element

ReqIF is persisted as XML, and therefore represents a tree structure. The top level element is called ReqIF and shown in Figure 4.2. It is little more than a container for a header (ReqIFHeader), a placeholder for tool-spefic data (ReqIFToolExtension) and the actual content (ReqIFContent). The content element is shown with all its details in Figure 4.3.

The ReqIF Content

The ReqIFContent has no attribute, but is simply a container for six elements. These are:

SpecObject A SpecObject represent an actual requirement. The values (AttributeValue) of the SpecObject depend on its SpecType.

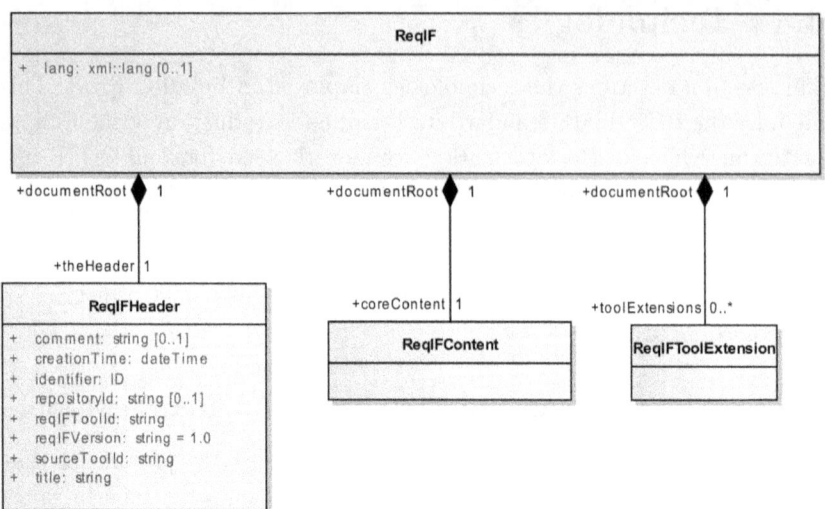

Figure 4.2: The top-level ReqIF element (from [OMG, 2011])

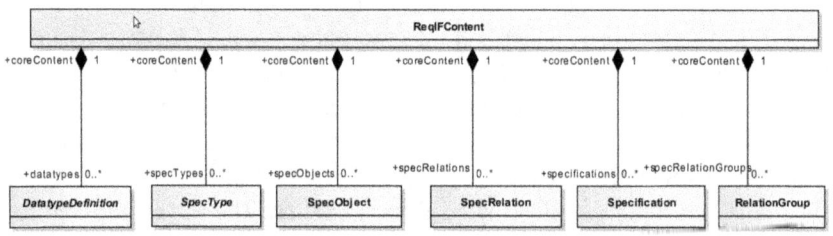

Figure 4.3: The ReqIFContent element (from [OMG, 2011])

SpecType A SpecType is a data structure that serves as the template for anything that has Attributes (e.g. a SpecObject). It contains a list of Attributes, which are named entities of a certain datatype and an optional default value. For example, a SpecObject of a certain type has a value for each of the SpecType's attributes.

DatatypeDefinition A DatatypeDefinition is an instance of one of the atomic data types that is configured to use. For instance, String is an atomic data type. A DatatypeDefinition for a String would have a name and the maximum length of the string. An attribute of a SpecType is associated with a DatatypeDefinition.

Specification SpecObjects can be grouped together in a tree structure

called Specification. A Specification references SpecObjects. Therefore it is possible for the same SpecObject to appear in multiple
Specifications, or multiple times in the same Specification.

In addition, a Specification itself may have a SpecType and therefore
AttributeValues.

SpecRelation A SpecRelation is a link between SpecObjects, it contains
a source and a target. In addition, a SpecRelation can have a
SpecType and therefore AttributeValues.

RelationGroup SpecRelations can be grouped together in a Relation-
Group, but only if the SpecRelations have the same source and target
Specifications.

This sounds strange, and in fact is a crutch brought in by one
vendor of requirements management tools, who was concerned about
compatibility with their products. This construct only makes sense
when considering some of the limitations of existing tools.

SpecElements and their Typing System

The previous section described the four element types that can have
attributes: SpecObjects, Specifications, SpecReleastions and Relation-
Groups. These four are all SpecElementsWithAttributes, or SpecElements
for short. This is shown in Figure 4.4. That figure also shows that each
has its own subclass of SpecType (SpecObjectType, SpecificationType,
SpecRelationType and RelationGroupType). A SpecType has any number of AttributeDefinitions, which ultimately determines the values of a
SpecElement. Correspondingly, a SpecElement can have any number of
AttributeValues. The AttributeValues of a SpecElement depend on the
AttributeDefinitions of the SpeElement's SpecType. This fact can not be
deducted from the model.

The AttributeDefinition references a DatatypeDefinition that ultimately determines the value of the AttributeValue of the corresponding
SpecElement. For each atomic data type of ReqIF, there is a corresponding DatatypeDefinition, AttributeDefinition and AttributeValue each. Before looking at the atomic data types, let's look at a concrete example.
Figure 4.5 shows the three classes for the atomic data type "String".

The actual AttributeDefinitionString has a type of DatatypeDefinitionString. This one happens to have one attribute, "maxLength". In
other words, one needs a concrete instance of DatatypeDefinitionString in
order to create an AttributeDefinitionString. An AttributeDefinition may

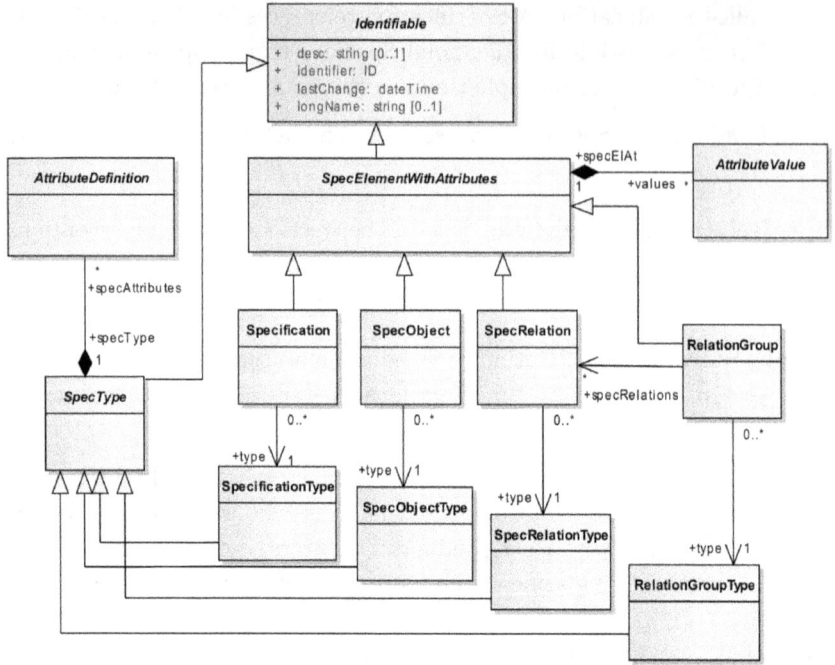

Figure 4.4: Elements with attributes and AttributeDefinitions (from [OMG, 2011])

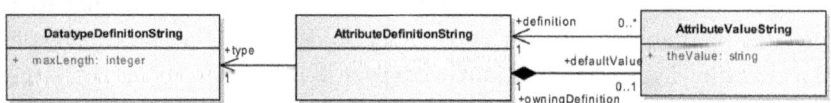

Figure 4.5: Elements with attributes and AttributeDefinitions (from [OMG, 2011])

also have a default value, which is of type AttributeValueString and may be null.

ReqIF supports the following atomic data types:

String A unicode text string. The maximum length can be set on the Datattype.

Boolean A boolean value. No customization is possible.

Integer An integer value. The maximum and minimum can be set on the Datattype.

Real An real value. The maximum and minimum can be set on the Datattype, as well as the accuracy.

Date A date- and timestamp value. No customization is possible.

Enumeration An enumeration Datatype consist of a number of enumeration values. The AttributeDefinition determines whether the values are single value or multiple value.

XHTML XHTML is used as a container for a number of more specific content types. The AttributeValue has a flag to indicate whether the value is simplified, which can be used if the tool used to edit only supports a simplified version of the content. For instance, if rich text is not supported, and therefore the new content is stored as plain text.

Other ReqIF Elements

ReqIF consist of 44 element types. The ones I just described are important for understanding ReqIF in general and this work in particular. Elements we omitted concern aspects like access control and identifier management.

Persistence

ReqIF provides a scheme for XML persistence. The following is an example of a small ReqIF file. This file contains one SpecType with one Attribute, and one SpecObject that uses it. It contains one Specification containing that one SpecObject.

```
1 <?xml version="1.0" encoding="UTF-8"?>
2 <REQ-IF xmlns:xsi="http://www.w3.org/2001/
    XMLSchema-instance" xmlns="http://www.omg.org/
    spec/ReqIF/20101201" xsi:schemaLocation="http
    ://www.omg.org/spec/ReqIF/20110401/reqif.xsd
    reqif.xsd http://www.w3.org/1999/xhtml driver.
    xsd">
3   <THE-HEADER>
4    <REQ-IF-HEADER>
5     <CREATION-TIME>2011-09-27T11
        :30:35.050+02:00</CREATION-TIME>
6     <SOURCE-TOOL-ID>ProR (http://pror.org)</
        SOURCE-TOOL-ID>
7    </REQ-IF-HEADER>
```

```
8    </THE-HEADER>
9    <CORE-CONTENT>
10     <REQ-IF-CONTENT>
11       <DATATYPES>
12         <DATATYPE-DEFINITION-STRING IDENTIFIER="
              _WkXXQejrEeC8vsU0vp6aHw" LONG-NAME="
              T_String32k" MAX-LENGTH="32000"/>
13       </DATATYPES>
14       <SPEC-TYPES>
15         <SPEC-OBJECT-TYPE IDENTIFIER="
              _WkY1Y0jrEeC8vsU0vp6aHw" LONG-NAME="
              Requirement Type">
16           <SPEC-ATTRIBUTES>
17             <ATTRIBUTE-DEFINITION-STRING
                  IDENTIFIER="_WkZzg0jrEeC8vsU0vp6aHw
                  " LONG-NAME="Description">
18               <TYPE>
19                 <DATATYPE-DEFINITION-STRING-REF>
                      _WkXXQejrEeC8vsU0vp6aHw</
                      DATATYPE-DEFINITION-STRING-REF>
20               </TYPE>
21             </ATTRIBUTE-DEFINITION-STRING>
22           </SPEC-ATTRIBUTES>
23         </SPEC-OBJECT-TYPE>
24       </SPEC-TYPES>
25       <SPEC-OBJECTS>
26         <SPEC-OBJECT IDENTIFIER="
              _Wkdd40jrEeC8vsU0vp6aHw">
27           <VALUES>
28             <ATTRIBUTE-VALUE-STRING THE-VALUE="
                  Test Requirement.">
29               <DEFINITION>
30                 <ATTRIBUTE-DEFINITION-STRING-REF>
                      _WkZzg0jrEeC8vsU0vp6aHw</
                      ATTRIBUTE-DEFINITION-STRING-REF
                      >
31               </DEFINITION>
32             </ATTRIBUTE-VALUE-STRING>
33           </VALUES>
34           <TYPE>
```

```
35              <SPEC-OBJECT-TYPE-REF>
                _WkYlYOjrEeC8vsUOvp6aHw</SPEC-
                OBJECT-TYPE-REF>
36            </TYPE>
37          </SPEC-OBJECT>
38        </SPEC-OBJECTS>
39        <SPECIFICATIONS>
40          <SPECIFICATION IDENTIFIER="_WkX-
                UOjrEeC8vsUOvp6aHw" LONG-NAME="
                Specification Document">
41            <CHILDREN>
42              <SPEC-HIERARCHY IDENTIFIER="
                _WkeE8OjrEeC8vsUOvp6aHw">
43                <OBJECT>
44                  <SPEC-OBJECT-REF>
                    _Wkdd4OjrEeC8vsUOvp6aHw</SPEC-
                    OBJECT-REF>
45                </OBJECT>
46              </SPEC-HIERARCHY>
47            </CHILDREN>
48          </SPECIFICATION>
49        </SPECIFICATIONS>
50      </REQ-IF-CONTENT>
51    </CORE-CONTENT>
52 </REQ-IF>
```

Getting persistence right was not trivial. EMF supports XML persistence, and EMF allows some tweaking of the XML. But creating the ReqIF persistence format was beyond of what could be achieved by tweaking.

In the end, the Verde team managed to get persistence to work, and there effort has been documented online[10].

Persistence outside the scope of my scientific work. Therefore, the details of the persistence mechanism will not be covered in detail. As far as ProR is concerned: The tool constructs an EMF-based model in memory, which is simply persisted by calling the code from the core to persist it as ReqIF on disk.

---

[10]http://nirmalsasidharan.wordpress.com/2011/07/29/dissecting-rifreqif-metamodel/

4.4.2 Eclipse

Eclipse[11] is a platform for general purpose applications with an extensible plug-in system. It is mainly known as an integrated development environment (IDE) for Java development, although the Java IDE is just one specialised application of the platform. Eclipse employs plug-ins in order to provide all of its functionality on top of the run-time system which is based on Equinox, an OSGi standard compliant implementation.

The Eclipse platform provides facilities for workspace management, GUI building, a help system, team support and more. These components consist of plug-ins. Plug-ins may provide extension points, to which other plug-ins may connect via extensions. A typical Eclipse installation contains hundreds of extensions.

ProR can run as a stand-alone Eclipse application, or it can be installed into any existing Eclipse installation, including Rodin.

There were at least two good reasons to use Eclipse for ProR. First, Rodin was an Eclipse application as well, which made integration of Rodin and ProR easy. This was one of the goals (see Section4.3). Second, I was deeply familiar with writing Eclipse-based applications, having co-authored a book on the Eclipse Rich Client Platform [Sippel et al., 2008]. Being familiar with Eclipse, I knew that the platform had the capabilities to support all the required features. Furthermore, EMF had the potential to simplify many aspects of the development.

4.4.3 Java

Eclipse is written in Java, and so are Eclipse applications. While Java is not the most elegant programing language in the world, it has the advantages of being proven in the field, fast and supported on many operating systems. The fact that I had over fifteen years experience in writing Java-applications made it attractive as well.

4.4.4 Eclipse Modeling Framework (EMF)

The Eclipse Modeling Framework [Budinsky et al., 2009] is a modelling and code generation facility. EMF provides tools and runtime support to produce Java code for the model and adaptor classes that enable viewing and command-based editing of the model.

EMF is attractive for ProR for a number of reasons:

---

[11]http://eclipse.org

- EMF can work off models described in XMI, which allows interoperability with other modelling tools. For instance, the digital representation of RIF could be used as the starting point for ProR, which sped things up considerably.

- EMF is modular. Halfway through the project, ProR switched to the EMF-based data model implementation from the ITEA-VERDE-Project (see Section 4.2). Thanks to the modular structure of EMF, this was straight forward.

- EMF takes care of many mundane tasks in GUI development.

- Rodin provides an EMF bridge. Even though Rodin is not based on EMF, there is a plug-in that provides an EMF-based bridge to the Rodin data model. This plug-in is actively maintained and makes it easy to integrate the data models from ProR and Rodin.

An EMF-application typically consists of three layers:

Model The model layer consists of the actual data structure and is stored in the form of the Ecore model. The Ecore Model can be either generated from scratch or imported (e.g. from an existing UML or XSD model). Customizations of the Ecore model include namespace, containment of elements (for persistence) and others. There is a corresponding Genmodel (Generator Model) that allows fine-tuning of the generated code for Model, Edit, Editor and Tests. Applied to the model layer, it generates the Java code for the data model.

Edit The Edit layer consists of so-called *ItemProviders*, which represent the bridge between the data model and a GUI. The ItemProviders can provide an alternative structure of the data. It is not unusual that the structure of the data model differs from the structure in the GUI. ItemProviders also provide basic information like labels and icons. They also collect the properties that are presented in the property view and collect the Commands that a user can perform on a data element. Last, they provide facilities to support Undo/Redo, Copy, Cut and Paste, Drag and Drop, and more. The ItemProvider code is also customised through an d generated by the Genmodel.

Editor EMF can also generate code for an Eclipse-based editor. Such an editor is generic, in the sense that it can be driven by any set of ItemProviders. The editor support many standard features that one would expect of a modern model editor, including support for Outline, Drag and Drop, Undo and Redo, etc.

EMF expects generated code to be modified. Annotations in the comments control which code is overwritten on regeneration.

Not many modifications of the model were required (in fact, the only modification at the time of this writing was the generation of the unique ID that is required for some model elements). However, it was planned to eventually extend the model with validators. EMF allows validators to "hook" directly into the model to validate certain properties of the model. This is quite useful for ReqIF, as there are a number of properties of valid ReqIF that cannot be validated on the XML level.

Most of our customisations take place on the ItemProviders. Our representation of the data in the GUI differs radically from the data model representation.

4.4.5   Modifying Generated EMF-Code

The code that EMF generates can be customised quite a bit by tweaking the *Generator Model*. The generator model stores information for the model generation. This includes mundane things like the directory where generated code should be stored or prefixes for the generated class names, as well as more sophisticated things, like the generation of notifications or whether the ItemProviders should be stateless or stateful.

It is recommended to configure as much as possible. But much desired behaviour can only be achieved by editing the code. As discussed in the previous section, EMF generates code that is intended to be modified. As the Verde project was responsible for the core code (including the data model elements), my preference was to not touch their code or their generator model. Fortunately, that was not necessary, mainly due to the clean architecture that EMF provides.

A central element of an EMF-based GUI are the so-called *Item-Providers*. They are used to adapt EMF objects, providing all of the interfaces that they need to be viewed or edited [Budinsky et al., 2009]. They represent the bridge between model and GUI, without any dependency to GUI code. There is an ItemProvider for every model element class. ItemProviders where fully in the scope of ProR, and I had to modify them heavily.

ItemProviders work their magic by combining a number of specialised providers. Code that needs one of those specialised providers uses the framework to adapt the model element to the required provider type. For example, a widely used pattern in GUI programming is the concept of a *LabelProvider*, which is used to find the label and icon to render for an object. Consider a control representing a tree structure (e.g. the Outline

View). To find the text strings and icons to render, the control consults the LabelProvider, requesting the label and icon for a given Object. Of course, the LabelProvider must be custom-written for the data model at hand. For a class to act as an EMF LabelProvider, it must implement the following interface:

```
1 public interface IItemLabelProvider
2 {
3    public String getText(Object object);
4    public Object getImage(Object object);
5 }
```

EMF generates the ItemProviders, which implement the IItemLabel-Provider and a number of additional providers (listed below). Generated methods can then be edited to change their behaviour (e.g. `getText()`). EMF generates default implementation that are typically "good enough" for testing. For instance, the default `getText()` implementation inspects the model element for an attribute called "name". Failing that, it looks for an attribute with the string "name" in the name.

An EMF ItemProvider implements the following provider interfaces:

IItemLabelProvider As just described, this provider is responsible for text label and icon. I modify this in several places to provide more dynamic behavior. For instance, the user can configure which attributes should be used as the label of SpecObjects.

ITreeItemContentProvider This provider controls how model elements are navigated. It provides a `getChildren()` and a `getParent()` method. This makes it possible to expose a data structure in the GUI that looks very different from the data structure of the underlying data model. ProR uses this frequently, as the ReqIF data model would be very hard to navigate and to read.

IItemPropertySource When an element is selected in the GUI, the property view typically shows the element's properties and allows them to be edited. This provider delivers a list of property descriptors, one for each property. The property descriptors contain not only the name and value of each property, but also type (text, multiple choice, etc.), category, whether it can be edited, and many more. I modify this heavily for elements that have dynamic properites, like a SpecObject. Each SpecObject can have a different set of attributes, and ProR shows all attributes in the property view.

IEditingDomainItemProvider The `EditingDomain` is a construct
that manages the command-based modification of objects. For instance, the context menu of any element in the GUI allows users
to perform certain operations that modify that element (e.g. by
adding a child element). The commands appearing in the GUI are
collected by consulting the EditingDomainItemProvider. I modify
this heavily, for example for typed elements (e.g. SpecObjects). I
add an additional command for each SpecType, allowing the user to
not only an untyped SpecObject, but also a typed SpecObject with
just one click.

Once a method is modified, it must be marked with a special annotation, otherwise it will be

4.4.6   The Standard Widget Toolkit (SWT) and JFace

The selection of Eclipse as the tooling platform suggest SWT/JFace as the
graphics toolkit. SWT is a low-level library of controls like buttons, labels,
and the like, while JFace is a higher-level library that provides components
like forms, tables, trees, including an MVC-based framework. Eclipse itself
is built on SWT and JFace, and EMF can generate an SWT/JFace-based
GUI.

SWT was developed as an alternative to the Swing toolkit, which is
part of the Java core. While Swing draws all components itself, SWT
relies on the operating system to render the controls. Both approaches
have strengths and weaknesses. The Swing-approach has the advantage
of a consistent look and feel across operating systems. It has the
disadvantages of sometimes not looking quite native, and by being slower.
The performance issues, however, have been addressed a long time ago
and are now negligible.

SWT, on the other hand, has a more native feel to it, because the
operating itself renders the controls. But as a result, the application's
behavior may differ from operating system to operating system, sometimes
significantly. In fact, this created a huge problem at one point in the
development, as described in Section 4.4.7.

Besides the technical differences between the two toolkits, there are
also architectural differences. While both toolkits discussed employ
the MVC design patterns, Swing arguably does so more elegant than
SWT/JFace. In turn, there are other toolkits for GUI building that
are much better than both, Swing and SWT/JFace, but not for the
Java/Eclipse ecosystem. Specifically, a web-based toolkit like GWT was

considered as well. This could be used by rendering a browser component inside an Eclipse Editor. This would have had the additional advantage that the application could have been extended with a web-based viewer and/or editor.

In the end, the decision was made to go with SWT/Swing, with the addition of Agilegrid (Section 4.4.7). While other toolkits could have potentially paid off in the long run, none of our requirements justified the increased initial investment. In addition, picking a conventional approach avoided the risk of encountering unexpected problems halfway through the implementation (and as will be described in the next section, even with a conventional approach and many years of experience, this kind of problem was encountered).

4.4.7 Agile Grid

Using the SWT/JFace code that EMF generated as the foundation resulted in a working system fairly quickly. I used the SWT/JFace GUI generated by EMF as the starting point, which not only gave me a working GUI, but also the initial architecture of the GUI. Specifically, features like property view, outline view, context menus, drag & drop, etc. worked out of the box. I still had to invest a significant amount of work to adjust the Editor view. This is the central place where users do their work, and this has to be done right for any tool to gain acceptance.

Unfortunately, the behaviour of SWT components can differ from operating system to operating system. The behaviour of the JFace TreeTable differed in such a way to disqualify it for our purpose. A TreeTable is a tree structure with collapsible elements, and multiple columns for each element. Figure 4.6 shows the JFace TreeTable, customised for ProR. This worked quite well to present the Specifications to the user.

This approach worked well — at least on Linux. It did not, however, work correctly on Windows. The JFace TreeTable control had a severe bug on Windows systems. The Linux-based control could adjust the height of each row in the table individually. This is important for ProR, as one cell may contain a long text that breaks over multiple lines, while a cell in the next row may only have a very short text with fewer lines. Consider the row height of REQ-1 and REQ-2 in Figure 4.6.

On Windows, all rows must have the same height. Using the same row height would either mean wasting screen real estate (by using the row height of the biggest row) or only showing the beginning of a value. This

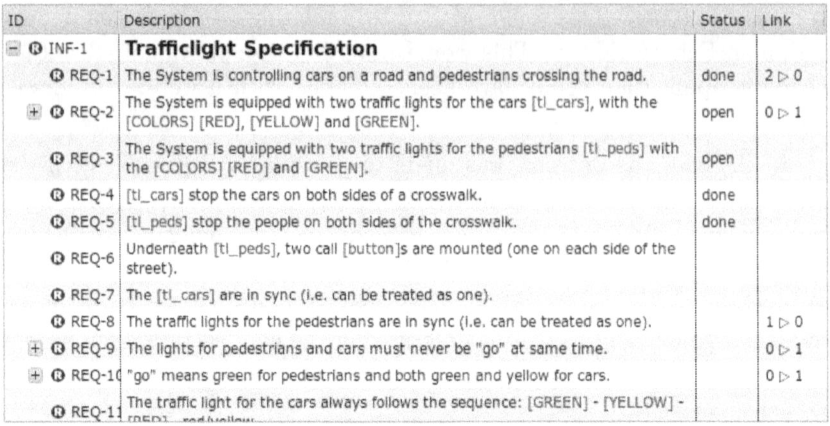

ID	Description	Status	Link
⊟ ❶ INF-1	**Trafficlight Specification**		
❶ REQ-1	The System is controlling cars on a road and pedestrians crossing the road.	done	2 ▷ 0
⊞ ❶ REQ-2	The System is equipped with two traffic lights for the cars [tl_cars], with the [COLORS] [RED], [YELLOW] and [GREEN].	open	0 ▷ 1
❶ REQ-3	The System is equipped with two traffic lights for the pedestrians [tl_peds] with the [COLORS] [RED] and [GREEN].	open	
❶ REQ-4	[tl_cars] stop the cars on both sides of a crosswalk.	done	
❶ REQ-5	[tl_peds] stop the people on both sides of the crosswalk.	done	
❶ REQ-6	Underneath [tl_peds], two call [button]s are mounted (one on each side of the street).		
❶ REQ-7	The [tl_cars] are in sync (i.e. can be treated as one).		
❶ REQ-8	The traffic lights for the pedestrians are in sync (i.e. can be treated as one).		1 ▷ 0
⊞ ❶ REQ-9	The lights for pedestrians and cars must never be "go" at same time.		0 ▷ 1
⊞ ❶ REQ-10	"go" means green for pedestrians and both green and yellow for cars.		0 ▷ 1
❶ REQ-11	The traffic light for the cars always follows the sequence: [GREEN] - [YELLOW] - [RED], red/yellow		

Figure 4.6: The SWT-based Specification View on Linux renders correctly.
On Windows, however, all rows have the same height.

was a known bug that had been around for many years[12] and there was
no reason to expect this to be resolved any time soon.

As I had no experience writing, let alone debugging Windows code, I
looked for alternative controls and eventually decided to use AgileGrid[13].

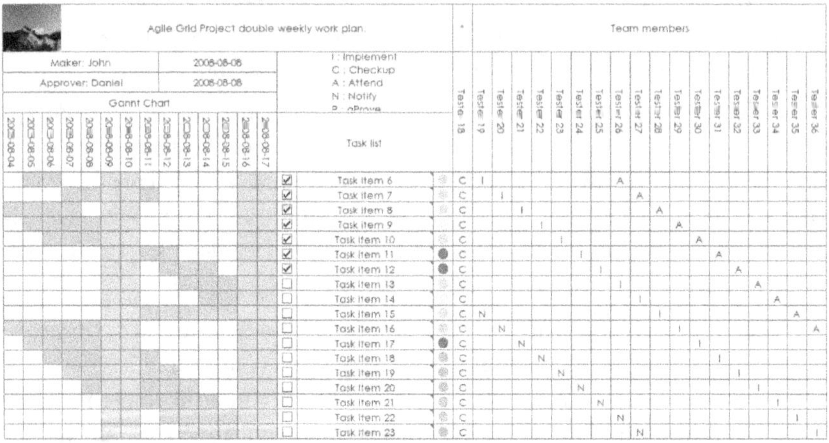

Figure 4.7: A demo that ships with Agile Grid to demonstrate its power

AgileGrid is an open source project that consists of a table control,

---

[12]https://bugs.eclipse.org/bugs/show_bug.cgi?id=148039
[13]http://agilegrid.sourceforge.net/

specifically for Eclipse. It is based on SWT, but the rendering of the table is not done by the operating system, but it is drawn directly on a canvas. Therefore, the appearance on all operating systems is identical.

AgileGrid scales well, which was one important criteria. It also allows individual row heights, which was — obviously — a core requirement.

AgileGrid does not support tree structures. While ProR has to render a tree structure, I did not see the need to allow nodes to collapse. I talked with potential users, and they confirmed that, while it would be nice to have, it wasn't a necessity. Thus, ProR now renders the tree structure fully expanded. The first column provides indenting, and hierarchical numbering in the margin communicates the tree structure to the user (Figure 4.8). This had an adverse effect on the performance, however. I had to create a content provider that translates the tree-structure into a flat table structure. The initial implementation was "dumb" in the sense that it did not do any caching, and required the traversal of the tree structure to find the position of objects. A better performing ContentProvider can be implemented anytime without an effect on the architecture.

In the end, the result was quite pleasing.

	Description	Status	Link
1	ℝ Demo of AgileGrid and ProR	true	
2	ℝ A second line at the same hierarchy	false	
3	ℝ Multiple Lines are now supported on all Operating Systems	true	
3.1	ℝ A child SpecObject	true	
3.2	ℝ Another Element on the same level	false	
4	ℝ Back to the highest hierarchy	true	
4.1	ℝ Some nesting		
4.1.1	ℝ ... and more nesting		
4.1.2	ℝ ... and even more	true	
4.1.3	ℝ Note the numbers in the left column.	false	
5	ℝ End of Demo!	true	

Figure 4.8: Agilegrid, adapted for ProR. Note that the first column is indented, and that the left margin indicates the hierarchy

SWT/JFace have the advantage that Eclipse is built on top of them, and that EMF can generate JFace Editor code. However,

4.4.8   The ReqIF Core

Initially, I built a partial EMF model based on the ReqIF specification, which at that time was in beta stage. This was quite useful for a number of reasons:

- It helped me to get a good understanding of ReqIF, as I had to deal with every single ReqIF feature that our software would support;

- It helped me to get a good understanding of EMF. EMF has a steep learning curve;

- It allowed me to focus on the essential GUI elements, as I built up the model element by element, while adding user interface elements at the same time.

At the same time, I was looking for other activities regarding RIF/ReqIF and encountered the Verde-project, which was driven mainly by the company itemis. While I was quite ahead with the development of the user interface, they were ahead regarding the RIF model. They, used the RIF 1.2 specification.

I briefly described my collaboration with Verde in Section 4.1.2. Switching out my half-finished ReqIF code with the Verde core was done within a few days. The bulk of the effort was due to the differences between RIF 1.2 and ReqIF. From that point of view it is kind of ironic that the decision was made later to build a ReqIF tool after all, and I had to go through the process once more in reverse.

For me, working with the Verde core was a good decision. It turned out that it was quite difficult to tweak EMF to write correct ReqIF. While EMF can — in theory — write arbitrary XML, the ReqIF XML is not very well suited for being generated by EMF. It took the Verde team a long time until all ReqIF features were persisted correctly, according to the specification.

Worth mentioning is the fact that RMF includes not only a ReqIF 1.0.1 core, but also a RIF 1.2 core and a RIF 1.1a core. Especially RIF 1.2. is, as of this writing, still in use. This allows, at least in theory, to create converters, or the ability to edit different formats. Code reuse in the GUI for this purpose was also explored for a while but turned out to not be feasible: this would have made the GUI code much more complicated, and there was no immediate value into doing this, certainly not from a scientific point of view.

4.5 Using ProR

This section will go through the more important features of ProR to provide an impression of the tool in action. I provided a similar introduction to the tool in [Jastram and Graf, 2011c]. I also created a screencast[14] that demonstrates the basic features of ProR. That screencast could be a useful complement to this section.

4.5.1 Installing ProR

ProR can be downloaded stand-alone, or installed into an existing application via its update site. The download is a convenient option for non-technical people who just want to get started with ProR. There are no special restriction for the update site version: ProR can be installed into any reasonably new Eclipse installation.

4.5.2 Creating a ReqIF Model

ReqIF models can be created in any Eclipse project, and manifest themselves as a .reqif file. Eclipse projects an have a *Nature*, which tailors the project towards a special purpose. ProR does not require a nature, which makes it easy to integrate ProR with other tools (while a project can have more than one nature, it can be cumbersome to make sure that all natures are properly set and configured).

A new ReqIF model can be created via the File | New... menu, where there is a wizard for a new "Reqif10 Model". The wizard will create a new ReqIF model with a very rudimentary structure, as shown in Figure 4.9. The model has one Datatype, one SpecType with one Attribute, using the Datatype, one Specification with one SpecObject that uses the SpecType.

ProR shows information in the outline and properties views. ProR provides a *Perspective*, which ensures that all relevant views are shown. As most Eclipse applications use these views, this is typically not an issue.

The editor in Figure 4.9 (the window in the middle) provides an overview of the model. The most important section is the one labelled "Specifications". Upon double clicking on one, it is opened in its own editor, as shown in Figure 4.10.

Each row represents a requirement (SpecObject), and each requirement can have an arbitrary number of attributes. Which specific attributes a requirement has depends on its type. We can see in the property view

---

[14]http://www.youtube.com/watch?v=sdfTNZduvZ4

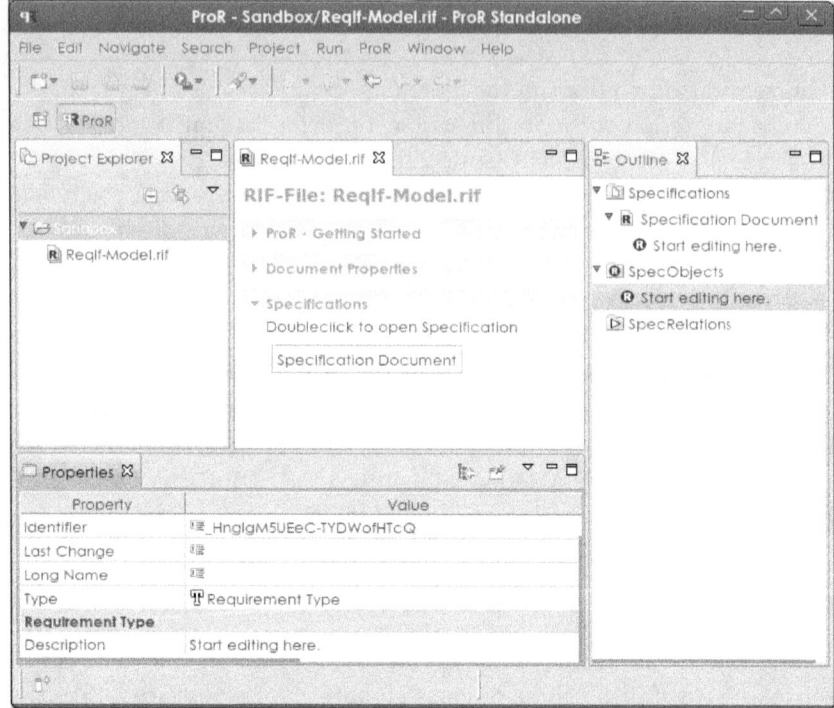

Figure 4.9: ProR with a newly created ReqIF model, as produced by the wizard

that the selected requirement is of type "Requirements Type", which has exactly one attribute called "description".

The editor can be configured to show an arbitrary number of columns. Each column has a name. If a requirement has an attribute of that name, then the value of that attribute is shown in the corresponding column.

To make the example more interesting, the following steps will be performed:

- Adding more attributes to the type "Requirements Type"

- Adding an additional column

- Enabling the *ID-Presentation*, a mechanism for automatically creating human-readable identifiers

- Adding of additional requirements

- Adding links between requirements

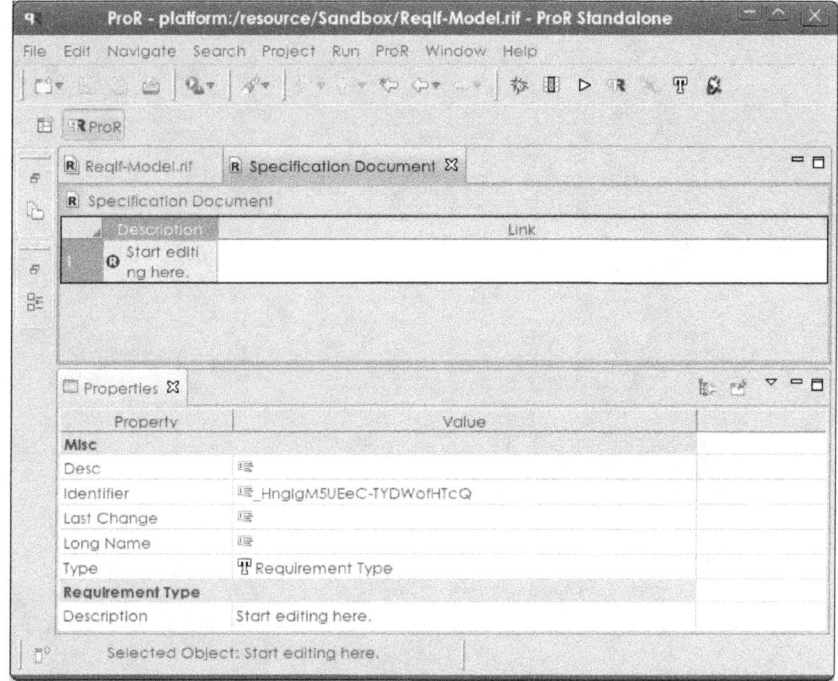

Figure 4.10: The Specification of the newly created ReqIF model

The result of these actions will resemble Figure 4.11.

4.5.3 New Attributes

The special dialogue for the datatypes is opened via PROR | DATATYPE
CONFIGURATION... or the corresponding icon in the toolbar (Figure 4.12).
The upper part of the dialogue shows the data structures, while the lower
part contains a property view that allows editing the properties of the
element that is selected in the upper part. New child or sibling elements
can be added via context menus.

Now two attributes to the type "Requirements Type" will be added:
an ID for a human readable identifier, and a status field, which is an
enumeration. The result is shown in Figure 4.12.

A new datatype for the ID called "T_ID" has been created. For the
status field, a new enumeration of type "T_Status" was created. In the
figure, one can see the properties of the selected element in the lower pane,
where they can be edited.

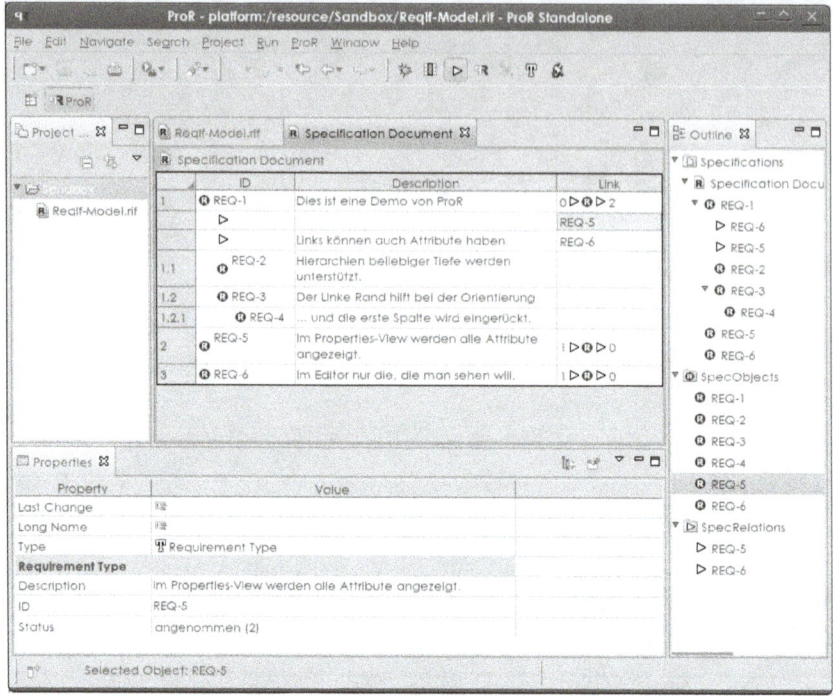

Figure 4.11: The Specification after adding some data

4.5.4   Configuration of the Editor

When closing the dialogue and select a requirement, the three properties will be visible in the properties view, where they can be edited. But the main pane of the editor still only shows one column. One can add new columns via PROR | COLUMN CONFIGURATION... (or the corresponding tool bar icon), which opens a dialogue for this purpose. The dialogue looks and works similar to the one for the data types. One more column called "ID" can then be added. The dialogue also allows the reordering of columns via drag and drop, and this mechanism is used to make the ID column the first one.

4.5.5   Generating IDs

The ID column in now visible in the specification editor, but it is empty. While IDs could simply be added by hand, this is error prone, and one would expect the tool to be able to handle this. ProR does not have the

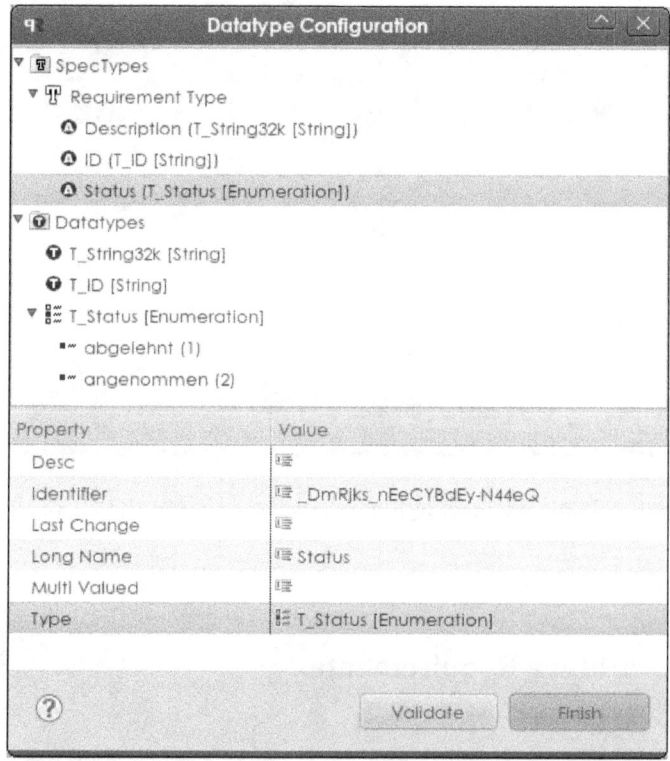

Figure 4.12: The datatype dialogue after adding some data

ability to generate IDs, but a "Presentation" can. Presentations are ProR-specific plug-ins that can modify the presentation of data and inspect and modify the data. Presentations will be described from a technical point of view in Section 4.6.

To add a presentation, the presentation dialogue is opened via PROR | PRESENTATION CONFIGURATION... (or the tool bar). The "Select Action..." dropdown lists all installed presentations, and selecting "ID Presentation" creates a new element. In the properties the prefix and counter of the presentation can be modified. But more important is the data type that is associated with the presentation. In this example "T_ID" shall be selected — and this is the reason why a new data type for the IDs was created earlier. The dialogue should now look like Figure 4.13.

After closing the dialogue, all requirements that did not have an ID yet will have received one by the presentation.

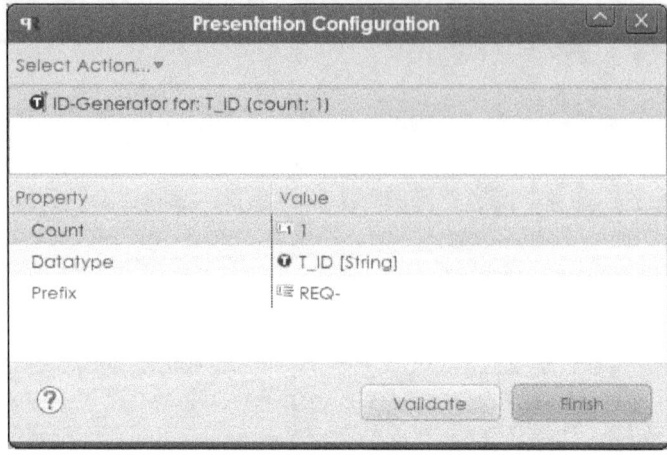

Figure 4.13: The presentation dialogue after creating the ID presentation

4.5.6   Adding Requirements

Finally everything is ready to add some data. This is mainly done via the context menus, but in several places, keyboard shortcuts are available. Upon opening the context menu for a requirement, new elements can be added via the "New Sibling" and "New Child" submenus. A specification is a tree structure of arbitrary depth, and the left margin indicates via a corresponding numbering scheme the position in the hierarchy. In addition, the left margin of the first column is indented (see Figure 4.8).

The context menu allows the creation of typed requirements — there is one entry for each user-defined type — which can save a lot of clicking. But it is also possible to add untyped requirements or even empty placeholders (*SpecHierarchies*). Adding a placeholder can be useful for referencing an existing requirement. Requirements may appear multiple times, both in the same specification and in other specifications of the same ReqIF model.

To allow the rapid addition of requirements, ProR provides the Ctrl-Enter keyboard shortcut. the new requirement is inserted below the one that is currently selected and has the same type.

Last, requirements can be rearranged. This can be done via drag & drop or copy & paste.

4.5.7 Linking Requirements

Requirements can be linked via drag & drop. As drag & drop is also used for rearranging requirements, it has to be combined with a keyboard modifier. The key that needs to be pressed is dependent of the operating system and is the same that is used for creating file links and the like.

Once a link has been created, the last column of the specification editor shows the number of incoming and outgoing links. It is possible to show the actual link objects (*SpecRelations*) via PROR | SPECRELATIONS..., which are then shown below the originating requirement. The last column of link objects shows the destination object (selecting that column will show the target requirement's properties in the property view).

Link objects can also by typed, resulting in them having attribute values as well. The values will be shown in the specification editor, if the columns are configured correspondingly.

This concludes the brief overview of the usage of ProR.

4.6 Extending ProR

While it is useful to have a tool for capturing structured requirements, the real value of ProR lies in its extendability and the integration with existing tool chains. To a degree, this is already achieved by supporting the ReqIF standard, which will provide an interface to commercial requirements engineering tools, once they support it[15]. But thanks to Eclipse, quite a bit more is available.

ProR exposes an extension point, which is a hook for other Eclipse plug-ins. *Presentations* can integrate themselves into the GUI and the model. A presentation consists of a service class and a data element that are connected through the extension point.

The service class offers many options for customising the GUI and to provide an integration with other Eclipse-based tools. The service class must implement the interface *PresentationService*. Rather than implementing the interface from scratch, one can override the class *AbstractPresentationService* which contains default implementations for all methods.

The data element allows the presentation to store data inside the ReqIF model. ReqIF uses the parent element *ReqIfToolExtension* for that purpose, which presentations can use as they see fit.

---

[15]Several commercial tools already support RIF 1.1a and RIF 1.2. As RMF contains cores for these standards, it would be fairly easy to build an interface using these standards.

A presentation is always associated with a *DatatypeDefinition*. The underlying reason will be clear shortly.

What exactly can a presentation do? Let's look at three simple presentations that exploit the extension point in various ways, and will then explore who they are implemented.

Headline Presentation ProR does not support formatted text yet. The headline presentation is a quick and dirty approach to provide some formatting in the meantime. This presentation allows the association of a data type with a bigger, boldface font. All attributes using that data type are rendered in that font.

ID Presentation The ID presentation was already presented in Section 4.5.5. While ReqIf requires IDs that ProR already generates, they are not particularly readable. This presentation generates human readable IDs with a configurable prefix.

RodinPresentation Rodin is a tool for formal modelling that was introduced in Section 4.3.1. This presentation allows the association model elements with requirements (e.g. a variable with a corresponding definition in the requirements document). The association can be established manually via drag & drop, and the model elements are subsequently visible in the requirements document.

Presentations are realised as Eclipse plug-in projects. This plug-in must provide an extension for the extension point *org.eclipse.rmf.pror.reqif10.presentation.service.presentation*, and it must extend the EMF model element *ProrPresentationConfiguration*. Everything else is implemented in the service class that implements *PresentationService*. In the following, a few selected methods of that interface will be presented.

```
IProrCellRenderer getCellRenderer(AttributeValue av);
```

This method allows us to provide an alternative *CellRenderer* (in order to use the default cell renderer, simply return null). The headline presentation overrides this method to provide a renderer that uses a different font. ProR ensures that the AttributeValue that is handed to this method is using the DatatypeDefinition that the headline configuration is associated with.

Next, let's have a closer look at the HeadlineConfiguration. Figure 4.14 shows the relevant elements from the model.

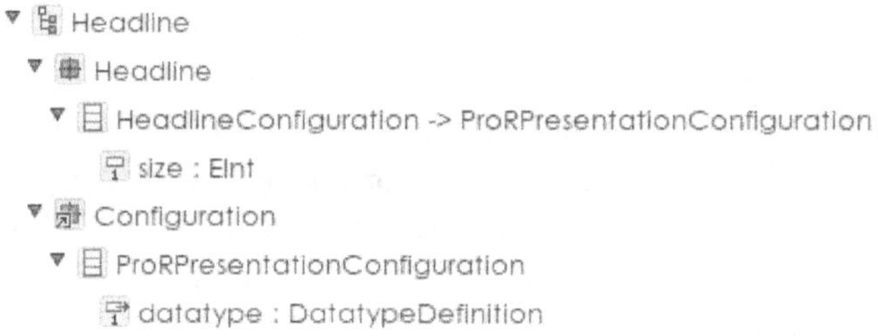

▼ ⬡ Headline
  ▼ ⬢ Headline
    ▼ ▤ HeadlineConfiguration -> ProRPresentationConfiguration
        ⬚ size : EInt
  ▼ ⬗ Configuration
    ▼ ▤ ProRPresentationConfiguration
        ⬚ datatype : DatatypeDefinition

Figure 4.14: The extension of the ProR model to accommodate data for the headline presentation

One can see that the model element HeadlineConfiguration has a new attribute called "size". As HeadlineConfiguration is derived from ProRPresentationConfiguration (also shown), it also has the attribute "datatype", which is used to identify the AttributeValues to which the presentation shall be applied.

Let's look at some more methods:

```
CellEditor getCellEditor(AgileGrid agileGrid,
        EditingDomain editingDomain, AttributeValue av);
```

Similar to the *CellRenderer*, the default *CellEditor* can also be replaced. For instance, the Verde-team hooked an XText-based editor into ProR, which provided syntax highlighting, auto-completion and similar features while editing [Jastram and Graf, 2011d]. As with the renderer, returning null will trigger the default editor.

```
public void openReqIf(ReqIf reqif)
```

This method allows a presentation to hook itself into the ReqIF model itself upon opening. It can then use all EMF mechanisms to do its magic. The ID presentation uses this feature to listen for newly inserted element, in order to set IDs if appropriate.

```
public Command handleDragAndDrop(Collection<?> source,
        Object target, EditingDomain editingDomain,
        int operation);
```

For a truly intuitive integration with other tools in an Eclipse environment, presentations can opt to serve drag & drop operations. If a presentation opts to do so, it must return an EMF command which will then be executed by ProR. The use of commands results in a seamless integration that supports undo, redo and similar behaviour.

This mechanism is used for the integration with Rodin. Rodin presents its model elements in a tree structure (for instance in the project view), from where they can simply be dragged into the ReqIF editor. Depending on the keyboard modifier, ProR links or integrates those elements into the specification. ProR only stores the internal ID of the referenced model element and retrieves the values from Rodin when they have to be rendered. This way, the content is never outdated.

4.7   Integration with Rodin

A key objective in the development of ProR was the ability to support the approach described in this work. The result is an integration plug-in that allows to add ProR functionality to Rodin[16]. Starting with Rodin 2.5, the update site for the integration is pre-installed, allowing the installation with a few clicks.

The integration allows the identification of phenomena within natural language requirements (Rodin already allows the identification of phenomena in formal model artefacts); It supports the creation of traces between arbitrary artefacts; and it tracks changes, marking traces as "suspect" if source or target of the artefact changes (allowing the re-validation of traces).

ProR supports the classification of artefacts as R, W, S and D "out of the box" by configuring an enumeration attribute for this purpose. Further integration is achieved by three features, provided by the integration plug-in.

Tracing phenomena   Tracing of phenomena is supported by colour-highlighting those words that correspond to phenomena in the requirements text. The user has to mark phenomena with square brackets. If a phenomena has been formally declared in the corresponding model, the phenomenon is rendered in blue, otherwise in red, reminding the user that an undeclared phenomenon is used. Further, if a word is encountered that is the designation of a phenomenon, but not marked as such (with square brackets), it is underlined red to remind the user that this word may represent an untraced phenomenon.

---

[16]The integration is documented at http://wiki.event-b.org/index.php/ProR

Creating Justification Traces Traces between artefacts can be created via drag and drop. This includes traces between informal artefacts, as well as informal and formal ones. Traces are established via drag and drop. The resulting traces and corresponding model elements can then be visualised inside the requirements specification. The model elements are referenced and are therefore always up to date. Traces can be annotated with arbitrary information.

Tracking Changes If the source or target of a trace changes, then the trace is marked as suspect by showing a small icon in a dedicated column. Two columns exist for source and target of the trace, respectively. By double-clicking on a cell, the user can reset the suspect flag after re-validating the trace.

There are still some limitations, however. While all required data structures exist, the tool would benefit from more sophisticated reporting, in particular with respect to the properties listed in Section 3.6.

The integration plug-in already added value during the creation of the case study (Chapter 5). The ability to trace phenomena and the immediate feedback from the colour highlighting helped to keep the terminology across artefacts consistent.

The ability to quickly identify suspect traces was also very useful. After an iteration of development, it allowed to systematically inspect the traces and to perform a manual validation where needed.

The creation of traces between formal and informal artefacts was not as seamless as hoped. The drag operation had to start in the outline of the formal model. It would have been nicer if it could have been initiated directly from the Rodin editor, but this would have required a change to the editor code. By using the project outline to initiate the drag operation, no Rodin code had to be modified to support the integration.

Further, currently it is only possible to create traces to events, but not to model elements within events (guards, actions, witnesses, parameters). This can be compensated by tracing to the event and adding relevant information to an annotation of the trace. With respect to the ProR approach, this is not problematic, as tracing to the complete event just inflates the satisfaction base slightly (see Section 3.2.3).

4.8 Conclusion

With ProR, academic research could be combined with the development of a tool that has industrial relevance. Arguable, the scientific value of ProR in itself is not particularly high: The tool provides a user interface for a data model that was provided by an international standard. Implementing such a tool is straight forward, in principle.

Nevertheless, the development of ProR represents a significant contribution in more than one way. First, the scientific value got established by the development of the Rodin integration and the support of the ProR approach. Second, the first Open Source implementation of the ReqIF data model, ProR provides a solid platform for other researchers in the field of requirements engineering. For instance, ProR has been used for the ITEA-project Verde[17] (Validation-driven design for component-based architectures) [Jastram and Graf, 2011d]. Third, by becoming an Eclipse Foundation project, the survival chances of ProR beyond this academic work were significantly improved. Not only increased this the visibility of ProR, but also helped building a community of ProR users and developers. And last, the significance of ProR in industry has been recognised as well. The Eclipse proposal attracted support from 13 "interested parties", which include Airbus as a supporter. While at the time of writing ProR is not yet used in production, there is significant commercial interest, which the firm Formal Mind (Section 1.8) is trying to serve.

At this time, ProR development continues with five Eclipse committers. Further, it is planned to continue academic work on the Rodin integration during the lifetime of the FP7 ITC project Advance[18].

---

[17]http://www.itea-verde.org/
[18]http://www.advance-ict.eu/

Chapter 5

A Case Study

Chapter 3 presented an approach and Chapter 4 described the tool ProR. In this chapter, a formal specification of a simple traffic light controller will be developed. This example uses Problem Frames to structure the problem, and the model will be built using the Event-B formalism. Customised tool support in the form of an integration of ProR and Rodin, described in Section 4.7 is used to support this approach.

This example has been used before [Jastram et al., 2010, Jastram et al., 2011]. This example is simple enough to understand, but complex enough to be interesting. Further, the example concerns state (which is modelled formally) as well as real-time (which is specified informally, as well as using temporal logic), demonstrating the mixing of formal and informal modelling elements.

A case study of a lift controller, using the ProR approach, is available in [Hallerstede et al., 2012].

5.1 The Goal: Crossing the Street

The goal of the system can be expressed in one sentence:

"A system that allows pedestrians to cross a road safely."

Note that this goal leaves a lot of questions open, and does not make any assumption regarding the solution. Open questions include:

- What kind of street is it? One lane or six lanes?

- Is there already some kind of infrastructure in place, like electricity, existing traffic lights, etc.

121

- Is it okay to stop traffic? For instance, this is not the case on a highway.

- In which country does this take place? What kind of regulations must be observed?

- How much space is available for providing a solution to the crossing-problem?

Many people will already have an idea on how to solve the problem. However, it is not wise to decide on a solution until the problem is fully understood. Here are some examples of quite diverse possible solutions:

- Build a bridge for the pedestrians

- Put the road underground

- Build a zebra crossing

5.2   Iteration 0: Elicitation of Requirements

The first step in describing the problem is the elicitation of requirements and the assumptions about the domain. Elicitation is outside the scope of this work. Methods for elicitation include introspection, interviews and analysis [Goguen and Linde, 1993]. For the street crossing problem, the starting point is the list of (simulated) requirements below, which is the result of the requirements elicitation phase. This information has been captured with ProR. This list is accompanied by a sketch that depicts the physical environment (Figure 5.1).

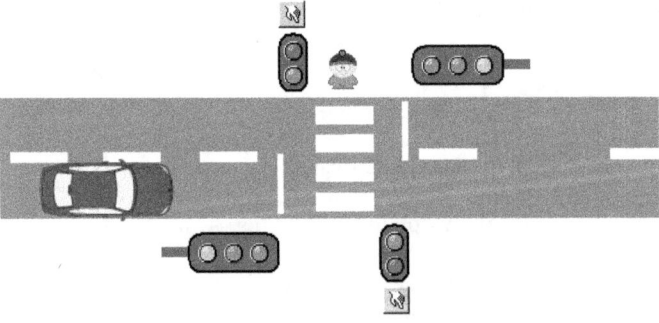

Figure 5.1: A sketch of the actual road and traffic lights

A0.1	The system allows pedestrians to cross the street safely
A0.2	The road is equipped with two traffic lights for the cars (colors red, yellow and green), one in each direction.
A0.3	The road is equipped with two traffic lights for the pedestrians (colors red and green), one on each side of the street.
A0.4	The traffic lights for the pedestrians are equipped with push buttons.
A0.5	The traffic for cars is usually green.
A0.6	Pedestrians can request permission to cross the street by pushing the push button.
A0.7	Pedestrians will get permission to cross the street t_1 seconds after the push button got pressed.
A0.8	The duration of the green light for pedestrians is t_2 seconds.
A0.9	The traffic light system follows the regulations for traffic lights of Germany (Richtlinien für Signalanlagen, RiLSA).

Closer inspection of these artefacts reveals that the list not only contains requirements, but also domain properties. It is also implied that the system will be used to control a traffic light system (and not, say, a pedestrian bridge), as some artefacts specifically refer to traffic lights and colours (e.g. A0.2 and A0.3, etc.). It is not clear, however, on whether the traffic lights are given, and therefore part of the domain W, or whether somebody already proposed a solution, implying that these artefacts belong to S. In other words, the system boundaries are not clear, the description is vague about what is to be constructed ("the system").

In this example, the objective to construct a controller, and the hardware is considered part of the domain. This is clarified in the next section, using the Problem Frames approach, which was introduced in Section 3.5.1.

These requirements have a number of additional weaknesses. For instance, the regulation cited in A0.9 requires a delay between the pedestrian light turning red and the light for the cars turning green. This delay depends on the width of the street, and this domain property is missing.

There are more issues, and some of them will become apparent only after modelling the system.

In the following, the **ProR approach** is used to identify and address these and other weaknesses, while building up the specification, consisting of formal and non-formal artefacts.

5.3   Iteration 1: The Problem Diagram

A problem diagram is an extension of a context diagram by adding requirements. It serves as a starting point for problem analysis. Realistic problems must be decomposed in a set of subproblems.

To construct the initial problem diagram, the domains and their shared phenomena must be identified. This is typically done by analysing the exiting requirements and is described in [Jackson, 2001]. A good starting point are the nouns found in the requirements text. The following list shows the identified domains and introduces designations:

System. The system has to be constructed, and it controls the traffic lights [tl_cars] and [tl_peds] via control signals.

Street. The [street] itself does not control any phenomena. The use of the synonym [road] will be discontinued.

Pedestrians. The [pedestrians] control their movement across the [street], which is modelled as [crossing], [stopping] and [waiting]. They also control the phenomenon [push].

Cars. Corresponding to [pedestrians], [cars] control the movement vehicles across the street, which is modelled as [crossing], [stopping] and [waiting].

Traffic light car. The [tl_cars] controls its lights, which are [red], [yellow] and [green].

Traffic light pedestrians. The [tl_peds] controls its lights, which are [red] and [green].

Push buttons. The [buttons] control their state

Just by creating this list (which is similar in nature to a glossary), a synonym was identified. The quality of the system description will be improved by removing it.

The corresponding problem diagram is shown in Figure 5.2. A problem diagram consists of exactly one machine domain (the controller) and an arbitrary number of domains (there are designed and given domains), as well as phenomena being shared between domains.

The problem diagram shows the domains and whether they share phenomena. The phenomena are labelled with the entity controlling it and the possible states (e.g. *red*, *green*), as described in the list above.

Problem Frames are only used to support this approach, and the problem diagram is merely used as an aid to structure our requirements. In particular, the problem diagrams shown here are not integrated into ProR.

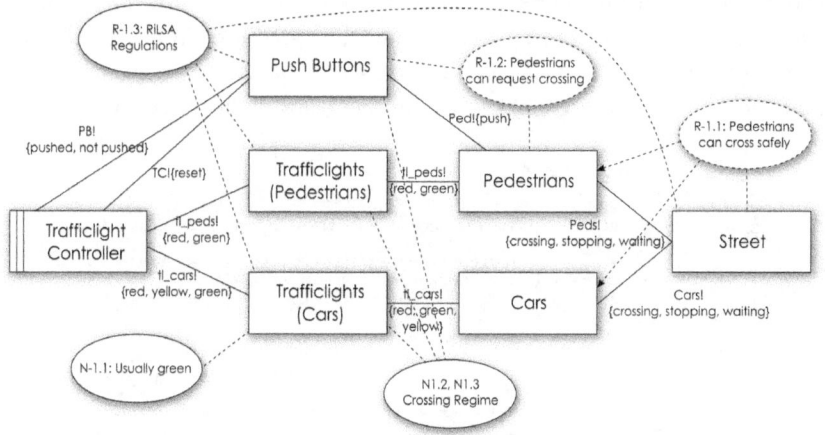

Figure 5.2: The initial problem diagram

The problem diagram represents the foundation for classifying the artefacts. In particular, it helps to clarify the system boundaries. The machine domain "Trafficlight Controller" in the problem diagram represents the system to be constructed. Therefore, all other domains are part of the environment.

From the problem diagram, it is immediately clear which phenomena belong to the environment (e) and which ones to the system (s): All phenomena that are not connected to the controller belong to e_h. Of the remaining phenomena, those that are controlled by the system belong to s_v, and those that are controlled by the environment (but visible to the controller) belong to e_v.

The problem diagram uses a simplification, where the phenomena [tl_peds] and tl_cars are used twice (once representing the control signals, and once representing the light visible to cars and pedestrians). This simplification is allowed [Jackson, 2001], and the phenomena are modelled as s_v.

Both states and values, are modelled as phenomena. For example, [tl_peds] is the phenomenon representing the traffic light state, while [red], [$yellow$] and [$green$] are phenomena representing possible values for [tl_peds].

With this knowledge, artefacts can be rephrased, using the new designations, and they can be classified as R, W or N.

With the aid of the problem diagram, one can restructure and improve the requirements, according to the ProR approach. In addition to evolution, this means classifying and marking artefacts and phenomena. Figure 5.3 shows the restructured artefacts, as presented in ProR. All artefacts are classified as R, W or N, all phenomena are listed in a glossary, and the phenomena are marked by square brackets in the artefacts.

Iteration 1

R1.1	The system allows [peds] [moving] across the [street] safely
W1.1	Two synchronised traffic lights for cars [tl_cars] are located on the [street], according to Fig. 5.1.
W1.2	The state of the car traffic lights is represented by [tl_cars], which represents a subset of [red], [yellow] and [green], meaning that the corresponding light is on.
W1.3	Two synchronised traffic lights for pedestrians [tl_peds] are located on the [street], according to Fig. 5.1.
W1.4	The state of the pedestrian traffic lights is represented by [tl_peds], which represents a subset of [red] and [green], meaning that the corresponding light is on.
W1.5	Two buttons are mounted on the bases of the pedestrian traffic lights [tl_peds], according to Fig. 5.1, allowing [peds] to [push] them.
W1.6	[peds] can press any of the push [button]s to trigger a [push] event.
N1.1	The traffic lights for cars [tl_cars] are usually [green].
R1.2	[peds] signal their wish to cross the [street] by [push]ing one of the [button]s.
N1.2	Between [push]ing the button for the first time in a cycle and [tl_peds] allowing pedestrians to cross, at most [t_1] seconds must pass.
N1.3	Upon turning [green], [tl_peds] must stay green for [t_2] seconds, with a tolerance of 5%.
R1.3	The traffic light system follows the regulations for traffic lights of Germany (Richtlinien für Signalanlagen, RiLSA)

Phenomena

[red]	e_v: used as control signal value (also s_h, when used as visible colour)
[yellow]	e_v: used as control signal value (also s_h, when used as visible colour)
[green]	e_v: used as control signal value (also s_h, when used as visible colour)
[tl_cars]	s_v: control signal, any combination of [red], [yellow], [green]
[tl_peds]	s_v: control signal, any combination of [red], [green]
[moving]	e_h: movement across the street
[stopping]	e_h: movement across the street
[waiting]	e_h: movement across the street
[street]	e_h: The street which can be crossed by [peds] or [cars]
[peds]	e_h: Movement of Pedestrians, one of [moving] [stopping], [waiting]
[cars]	e_h: Movement of Cars, one of [moving] [stopping], [waiting]
[push]	e_h: Act of pushing the call button
[reset]	s_v: Act of resetting the call button
[button]	e_v: Status of call button signal (boolean).
[t_1]	e_v: duration
[t_2]	e_v: duration

Figure 5.3: The classified and restructured artefacts after iteration 1. This is a screenshot taken from ProR.

The marking of phenomena immediately creates a number of use-traces. This could be written as follows (e.g. for W1.4):

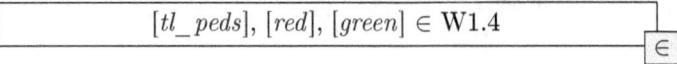

$$[tl\_peds], [red], [green] \in \text{W1.4}$$

Figure 5.3 visualises these relationship as well, by highlighting the used phenomena directly in the natural language text.

The evolution of artefacts can be inferred by comparing the initial artefacts with the revised artefacts (Figure 5.3. With such a small number, it is easy to recognise the evolution, e.g.

$$A0.2 \rightsquigarrow W1.1, W1.2$$

A more practical and scalable approach would be the creation of revisions. While ProR itself does not support this, it could be realised by using existing Eclipse plug-ins, for instance by integrating ProR with a version control system like Subversion or git. Users could then access the history of a ReqIF file and use the version control to create diff-views, showing the changes between revisions.

The evolution shown here should be validated by a domain expert.

At this point, all artefacts are still informal. Designations for the phenomena were already introduced, but this should not inhibit a stakeholder from understanding the artefacts. In fact, introducing designations and structuring the system description as shown in Figure 5.3.

Nevertheless, the restructuring already allows to check some properties of the system description that were described in Section 3.6. For instance, (3.24) is violated, as [reset] is not used by any artefacts. This can be remedied by introducing a new requirement:

R1.4	The [button] is [reset], once the [pedestrians] have crossed the [street].

At what point the properties are validated is a matter of taste. It is also clear that adequacy (3.4) is not realised yet, as nothing has been specified yet — so far, only the problem, not the solution, has been described.

5.4   Iteration 2: A First Step to Formalisation

In this section, the first formalisation is created, following the outline depicted in Figure 3.4. Modelling the first requirement typically involves a lot of work, as a good part of the domain model must be built before the actual requirement can be modelled.

Before starting with the modelling process, it is worth contemplating what the purpose of the model is, and which elements —conceptually— shall be modelled and which not. The main purpose of creating the model is to eventually implement a high-quality implementation. Further, the model is a means of communication and used for managing change. It may be employed in project management or testing as well.

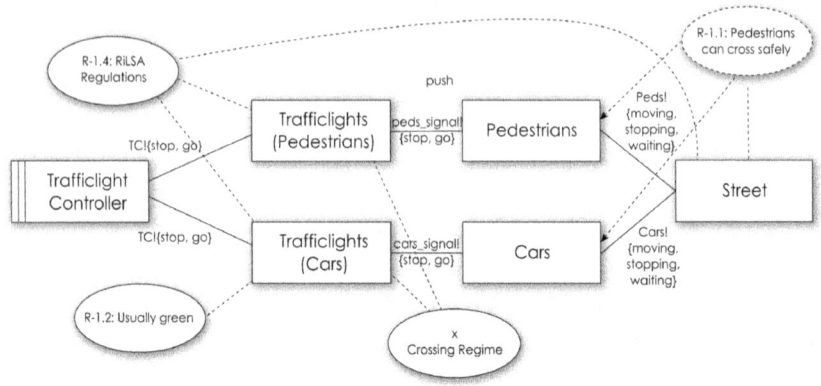

Figure 5.4: A modified subset of the initial problem diagram, reflecting elements in the initial formal model

5.4.1   Architectural considerations

The decision was made to build a model in Event-B. It is used to model the safety-critical aspects of the system, which is something that Event-B is well-suited for, as long as those aspects can be expressed as invariants, as was argued in Section 3.3.

The structure of the model has an impact on readability and extensibility. The main structuring feature of Event-B is refinement — how should it be employed? This case study employs Problem Frames for structuring the problem, and the problem frames structure can help planning the refinement. Specifically, Problem Frames work by superimposing sub-problems onto each other (see Section 3.5.1). Event-B refinement can be used to introduce new domains from the problem diagram in subsequent refinements. This technique is used in this example to introduce the '[button]s in Section 5.6.

To start the formal modelling of the system, a subset of the initial Problem Frames diagram (Figure 5.2) has been selected to be modelled initially, shown in Figure 5.4. The domain "Push Buttons" has been left out and will be superimposed later. Also, the phenomena for [tl_cars] and [tl_peds] were simplified into [cars_signal] and [peds_signal] to express the essential information that these signals represent in terms of [*stop*] and [*go*]. In Section 5.5, data refinement is used to create a mapping between the phenomena.

Working with time can be problematic in Event-B (see Section 3.3.2). There is no notion of real-time. Time can sometimes be modelled using counters, where each "tick" represents a fixed time length, as demonstrated in Section 5.7. Doing so allows the modelling of certain temporal properties, at the expense of complexity in the model. Whether this added complexity is justified depends on the project's requirements. With the ProR approach, it is always an option to keep artefacts informal.

Another consideration is whether the model should be used to generate actual code, eventually. An Event-B model used for code generation must exhibit certain properties (all non-determinism has to be removed from the model, for instance) [Edmunds and Butler, 2010]. This is not a consideration for this case study.

5.4.2 The First Requirement

Figure 3.4 depicts the work flow for building the formal model iteratively. Let's follow that process and select the first requirement:

R1.1   The system allows [*peds*] [*moving*] across the [*street*] safely

---

This requirement is central to the system, it is safety-critical and therefore should be modelled formally. At the same time, it is missing some properties that a "good" requirement should have [Hood and Wiebel, 2005]. Specifically, there is no clear criteria telling whether "safety" has been achieved, which has some non-functional aspects to it. In fact, R1.1 would be better classified as goal (Section 2.6) than a requirement. Techniques for developing requirements from goals exist (e.g. [Van Lamsweerde et al., 2001]) and are outside the scope of this work. In the following, R1.1 will be evolved in a functional and a non-functional component.

The functional aspect of "safety" concerns crossing the street, without being run over by a car. This can be expressed simply by stating the desired behaviour of cars and pedestrians, with respect to the street, as follows:

R2.1   When [*peds*] are [*moving*] or [*stopping*] on the [*street*], [*cars*] must be [*waiting*].

This requirement still reflects the relationships shown in Figure 5.4

(i.e., the traffic lights are not mentioned in the requirement). Unfortunately, R2.1 is weaker than R1.1. The other safety-related requirements will be captured with the following non-functional requirement:

N2.1	The system has additional safety properties.

As the focus of this work is the formal modelling of the functional aspects of the system, N2.1 is merely there to capture the non-functional aspects of R1.1. In practice, it would be much more elaborate. Nevertheless, N2.1 is now part of the system description, has its own traceability and has to be realised, if the system description is to be considered consistent.

The evolution of R1.1 can now be documented as:

R1.1 ⤳ R2.1, N2.1
⤳

5.4.3   Formalisation

The new requirement R2.1 is now fit to be formalised. The ProR approach requires the identification and modelling of phenomena. There is no "right" or "wrong" here — there are many ways for building a model. Consider the street: Should it be part of the model, and if so, which phenomena does it share? While the diagram in Figure 3.5 shows the street as a domain, this does not meant that it has to be modelled formally. For this case study the decision has been made to exclude the street from the formal model. The formal model will reflect the car's and pedestrian's movement, but the fact that this movement happens across the street. This is captured in Figure 5.1, which should be considered part of the system description as a domain property.

In the following, only relevant elements of the model are shown. The complete model is included in Appendix B.

The phenomena used by R2.1 are [peds] and [cars], which will be modelled in an Event-B context. Both can have the states [moving], [stopping] and [waiting]. The states will be modelled as constants in the Context ctx2. Event-B requires constants to be typed. Therefore, a carrier set [moving] is introduced. For brevity, only the declarations are omitted, leaving the following axiom:

axm1 $: partition(MOVING, \{moving\}, \{stopping\}, \{waiting\})$

Next, the phenomena [peds] and [cars] can be introduced as Event-B variables in the machine mac02a. Corresponding to the context, declarations are omitted.

w2.1 : $peds \in MOVING$

w2.2 : $cars \in MOVING$

These two invariants correspond to the descriptions of [peds] and [cars] in Figure 5.3. Those descriptions are domain properties that should be made explicit:

W2.1	Movement of [peds] is one of [moving], [stopping] or [waiting].
W2.2	Movement of [cars] is one of [moving], [stopping] or [waiting].

This, in turn, results in two equivalence traces:

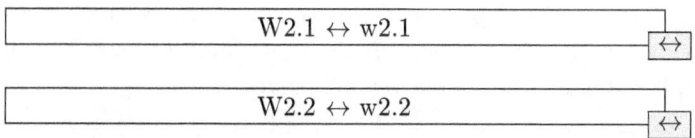

This is sufficient to formalise R2.1:

r2.1 : $(peds = moving \lor peds = stopping) \Rightarrow cars = waiting$

This is an equivalence:

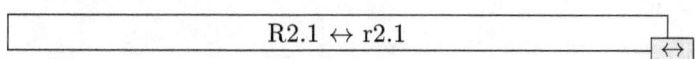

5.4.4 Completing the Machine

The model described so far has no ability to change state, and no initial state. Both are achieved by events.

The initialisation is performed by a special event "Initialisation", which must not result in an invariant violation. For this example, it is assumed that neither cars nor pedestrians are on the road, meaning that both phenomena are set to [waiting]. This is a domain property that must be documented. The initialisation can have multiple actions, therefore the action is traced, not the event.

EVENTS

Initialisation
   begin

        w2.3 : $peds := waiting$
        w2.4 : $cars := waiting$
   end

W2.3	Upon activating the system, [*peds*] are [*waiting*].
W2.4	Upon activating the system, [*cars*] are [*waiting*].

W2.3, W2.4 ↔ w2.3, w2.4	
	↔

Before events for state transitions of [*peds*] and [*cars*] can be provided, their behaviour must be specified as well. As already hinted at with W2.1 and W2.2, the real-world behaviour was reduced to three conceptual states. The model will provide events to cycle through these states in the order [*waiting*], [*moving*], [*stopping*]. The following shows the model for [*peds*], [*cars*] are modelled correspondingly:

EVENTS
Event *peds_ waiting_ to_ moving* $\hat{=}$
   when

        W2.1a : $peds = waiting$
   then

        W2.1b : $peds := moving$
   end
Event *peds_ moving_ to_ stopping* $\hat{=}$
   when

        W2.1c : $peds = moving$
   then

        W2.1d : $peds := stopping$
   end
Event *peds_ stopping_ to_ waiting* $\hat{=}$
   when

        W2.1e : $peds = stopping$
   then

        W2.1f : $peds := waiting$
   end

END

Introducing these events makes the trace between W2.1 and w2.1 suspect. In fact, this model represents a strengthening of W2.1. However, in Section 3.3.2 it was argued that the formal domain properties may only be weaker, not stronger than the informal ones. Thus, there is either a problem in the model or in the domain properties. A domain property corresponding to what has been modelled here would be:

W2.1'   [peds] that are not on the [street] are [waiting]. Upon entering the [street], they are [moving], followed by [stopping], before [waiting] again.

This domain property is written to already anticipate the reaction of pedestrians to traffic lights, even though the traffic lights have not yet been modelled.

This is less than optimal, and whether it is acceptable is up to the domain experts. While there may be a better way to model the behaviour of pedestrians, this behaviour is now adapted for this case study (and W2.2' correspondingly).

5.4.5   Proof Obligations

In its current form, the Event-B creates seven proof obligations, but only five are discharged. This is shown in Figure 5.5. This makes sense: So far, a part of the world has been modelled, as well as one requirement that describes how the world *should* behave, once the system is specified. But nothing has been specified yet.

The undischarged proof obligations can be used to identify the problematic artefacts by using traceability. This is trivial in this case. The undischarged proof obligations are *peds_waiting_to_moving/r2.1/INV* and *cars_waiting_to_moving/r2.1/INV*. These can be traced to R2.1 which may be violated, if [peds] start [moving], while [cars] are not [waiting] (or the other way around, respectively).

The system does not control pedestrians nor cars — it controls traffic lights. Therefore, first the behaviour of pedestrians and cars with respect to a traffic light must be modelled (domain property), which in turn allows the formal specification to be written such that R2.1 is realised and all proof obligations regarding r2.1 are discharged.

It is a matter of taste whether refinement shall be used or not. Specifically, r2.1 could be removed from the initial machine. A first refinement could then add the behaviour of the pedestrians and cars with respect to the traffic lights, while a second refinement could add r2.1 and the specification elements that realise it. On the other hand, all three machines could be rolled into one.

For this case study, the initial machine will be modified to introduce the traffic lights and the behaviour of cars and pedestrians with respect to the traffic lights. A first refinement will then add r2.1 and the corresponding specification elements. With this structure, the initial machine only contains domain properties, while the first refinement contains a requirement and its corresponding specification elements.

5.4.6   Modelling Traffic Lights

First, the phenomena representing the possible states of the traffic lights have to be added to the context. As explained earlier, initially the states are modelled as [stop] and [go]. Only later will this be changed to the actual traffic light colours using data refinement.

w2.5 : $partition(SIGNAL, \{stop\}, \{go\})$

This must be documented:

W2.5	Conceptually, the traffic lights [peds_signal] and [cars_signal] can indicate a [stop] or [go] signal, which is represented in the form of colours.
W2.6	The initial state for [peds_signal] and [cars_signal] is [stop]

Second, the machine will be modified by adding variables for the traffic lights, as well as events that modify it. The following shows the model elements for [peds_signal], the traffic light for cars is modelled correspondingly. Not shown is the initialisation, which sets both traffic lights to [stop]:

VARIABLES

peds_signal

INVARIANTS

w2.5a : $peds\_signal \in SIGNAL$

EVENTS

Event   set_peds_signal $\widehat{=}$

any

        signal
where

    w2.5c : $signal \in SIGNAL$
then

    w2.4d : $peds\_signal := signal$
end
END

Using the model yields the following traceability:

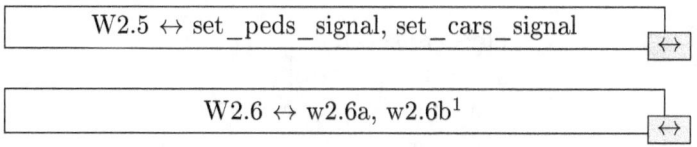

> W2.5 $\leftrightarrow$ set_peds_signal, set_cars_signal

> W2.6 $\leftrightarrow$ w2.6a, w2.6b[1]

Last, the behaviour of pedestrians and cars with respect to the signals must be modelled. This is undocumented domain knowledge that is so common that it is unlikely that a more traditional development approach would document it:

W2.7	[*peds*] start [*moving*] only if [*peds_signal*] is [*go*]. If [*peds_signal*] turns to [*stop*], [*peds*] that are [*moving*] are [*stopping*] and will be [*waiting*], once they finished crossing.
W2.8	[*cars*] start [*moving*] only if [*cars_signal*] is [*go*]. If [*cars_signal*] turns to [*stop*], [*cars*] that are [*moving*] are [*stopping*] and will be [*waiting*], once they finished crossing.

These domain properties can be partially realised adding guards to the events peds_waiting_to_moving and cars_waiting_to_moving. The guard would only the execution of the action if the corresponding traffic light indicates [*go*].

> W2.7 $\rightarrow$ w2.7[2], peds_waiting_to_moving,
> peds_moving_to_stopping,
> peds_stopping_to_waiting

> W2.8 $\rightarrow$ w2.8[3], cars_waiting_to_moving,
> cars_moving_to_stopping,
> cars_stopping_to_waiting

---

[1] Actions in event Initialisation
[2] Additional guard for event peds_waiting_to_moving

→

The trace includes the transition events as well, as the property relies on those events to realise the desired behaviour. This time, the trace is not an equivalence, as the formal domain property is weaker than the informal one, because it does not enforce the transition from [*moving*] to [*stopping*] upon the change of the signal. This is permissible, as described in Section 3.3.2.

This completes the modelling of the domain properties that are necessary for modelling and realising R2.1.

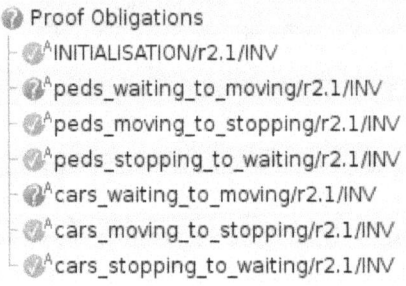

Figure 5.5: Proof obligations for mac02b

5.4.7 Realising Requirement R2.1

The requirement R2.1 has already been formalised and will be added to the machine mac02b, which is a refinement of mac02a. Doing so will, again, result in seven proof obligations, out of which two are not dischargeable (Figure 5.5), as discussed in Section 5.4.5.

The issue is that the system currently allows the pedestrians to start moving while the cars are still moving. This in turn is possible, because both signals can be set to "go" at the same time. The objective is to modify the behaviour of the system in a way to ensure that the invariant r2.1 is never violated. The naive approach (adding a guard to peds_waiting_to_moving) is not allowed, as the controller cannot directly constrain the pedestrians, as can be seen from Figure 5.2. This is also clear from the classification of [*peds*] as e_h. Assuming that the corresponding artefact is a specification element (S), this would also violate (3.25).

---

[3] Additional guard for event cars_waiting_to_moving

The system can only constrain the behaviour of the traffic lights, which means adding guards to set_peds_signal and set_cars_signal. One can add a guard that allows the pedestrian light only to change to "go" if the cars are waiting:

This can be prevented by adding guards grd1 and grd2 to the two events in question, as shown in the model further below.

But again, as is shown in Figure 5.2, the controller has no concept of the movement of pedestrians. And correspondingly, the proof obligations cannot be discharged: The system need to know how the traffic lights and pedestrians interact. This is an assumption that has to be made and can be stated as follows:

W2.9   When [peds_ signal] indicates [go], the [cars] are [waiting].

W2.10   When [cars_ signal] indicates [go], the [peds] are [waiting].

These properties can be modelled as invariants:

w2.9 : $peds\_signal = go \Rightarrow cars = waiting$

w2.10 : $cars\_signal = go \Rightarrow peds = waiting$

This is not enough, however, and undischarged proof obligations confirm this. Traceability indicates that this time the problem lies with the new domain properties, w2.9 and w2.10 with respect to the events peds_waiting_to_moving and cars_waiting_to_moving. Intuitively the problem is that both traffic lights can still switch to [go] while cars and pedestrians are waiting. A specification element can be added that prevents this:

S2.1   [tl_peds] and [tl_cars] must never be [go] at the same time.

S2.1 can be expressed as an invariant, as shown below. To prevent the invariant from being violated two more guards have to be added, resulting the the following additions to the model:

INVARIANTS

s2.1 : $\neg(peds\_signal = go \land cars\_signal = go)$

EVENTS

extends *set_peds_signal*
where

grd1 : $cars = waiting$

$$\texttt{s2.1a} : \neg(signal = go \wedge cars\_signal = go)$$
   end
 Event *set_ cars_ signal* $\mathrel{\widehat{=}}$
 extends *set_ cars_ signal*
    where

$$\texttt{grd2} : peds = waiting$$
$$\texttt{s2.1b} : \neg(peds\_signal = go \wedge signal = go)$$
    end

Section B.3 shows the complete machine model, as well as the updated list of requirements.

5.5   Iteration 3: Data Refinement

Before addressing the remaining requirements, the simplification regarding the traffic light colours must be taken care of. In Figure 5.2, the phenomena [*tl_ cars*] and [*tl_ peds*] were introduced, representing the colours of the traffic light. But in Figure 5.4, the signals were modelled as "stop" and "go". This will now be rectified by making the connection between the two. Section 3.3.2, described how consistency is maintained across refinement levels. This approach will now be applied here.

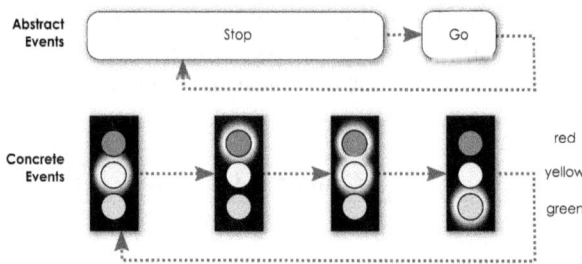

Figure 5.6: Data Refinement of the traffic light states for cars

Data refinement allows to state abstract properties in a concise way, while the implementation details are addressed later. This allows reasoning about some fundamental properties, as done here. Arguably, S2.1 would be more complicated if colours had been employed from the beginning.

There are other situations where this approach can be exploited: For product lines, some abstract properties could be realised in different concrete implementations. In this example, "stop" and "go" could be signaled with a barrier, as found in railroad crossings. A carefully crafted abstraction would therefore support the automated verification of different concrete implementations.

The relationship between "stop" and "go" and the colours, respectively, is a design decision. The informal notation does not have to be expressed in natural language — in this case it is more adequate to use a diagram, as shown in Figure 5.6. The notation does not follow a standard. Another option would have been to use something more precise like a UML state diagram. This is supported by UML-B [Snook and Butler, 2006], for instance. The validation of the diagram with a domain expert, however, is indispensable. This design decision can be recorded as follows:

D3.1   The relationship between [*tl_peds*] and [*peds_signal*] is a mapping of [*red*] to [*stop*] and [*green*] to [*go*], respectively.

D3.2   The relationship between [*tl_cars*] and [*cars_signal*] shall adhere to Figure 5.6.

Using colours requires the introduction of new constants (for the colours) and new variables for the traffic light states. Here is the definition of the colours, which are introduced in a context ctx03 that extends ctx02 from the previous iteration:

`colours` $: partition(COLOURS, \{red\}, \{yellow\}, \{green\})$

And below is the definition of the variables [*tl_peds*] and [*tl_cars*], which are added to the machine mac03a that refines mac02b. The connection between the colours and [*stop*] and [*go*] can be established with invariants that trace to D3.1 and D3.2.:

VARIABLES

    tl_peds

    tl_cars

INVARIANTS

    `d3.1a` $: tl\_peds \in \{\{red\}, \{green\}\}$

    `d3.2a` $: tl\_cars \in \{\{red\}, \{yellow\}, \{green\}, \{red, yellow\}\}$

    `d3.1b` $: peds\_signal = go \Leftrightarrow green \in tl\_peds$

    `d3.2b` $: cars\_signal = go \Leftrightarrow green \in tl\_cars$

This results in the following traceability:

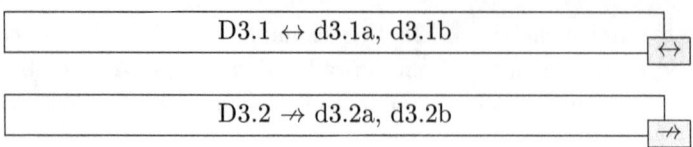

The formal model elements d3.1a and d3.1b are equivalent to D3.1, because only two states (and two state transitions) exist. In contrast, d3.2a and d3.2b are weaker than the design decision D3.2, as state transitions other then those shown in Figure 5.6 are permitted, in principle. For a design decisions. But as design decisions are on the right side of (3.4), they may not be weaker in the formal model (as discussed in Section 3.3.2).

To remedy this, the formal model refines the event set_cars_signal into four events, representing the state transitions:

- set_tl_cars_red_to_redyellow

- set_tl_cars_redyellow_to_green

- set_tl_cars_green_to_yellow

- set_tl_cars_yellow_to_red

The guards and actions of these events ensure that the state transitions specified by the events correspond to the state transitions in Figure 5.6. The updated traceability is:

> D3.2 ↔ d3.2a, d3.2b, set_tl_cars_red_to_redyellow,
> set_tl_cars_redyellow_to_green,
> set_tl_cars_green_to_yellow,
> set_tl_cars_yellow_to_red
> ↔

Once modelled, a number of additional adjustments to the model are necessary: Introducing the gluing invariants requires introducing witnesses in the corresponding events. All the additional model elements represent specification elements S that realise the design decisions D3.1 and D3.2. As the design decisions have been modelled formally, the theorem prover can guide the creation of of those specification elements. The resulting model can be found in Appendix B.4.

5.5.1 Temporal Logic

In its current form, Event-B does not generate proof obligations to validate that the state transitions are realised as shown in Figure 5.6.

Like with all other traces so far that bridge the formal and informal realm, the validation is done by reasoning. In this case, however, it is possible to articulate the state transitions in another formalism. This is demonstrated here by using linear temporal logic (LTL) [Plagge and Leuschel, 2010]. LTL can actually be understood as an extension to Event-B, complementing its standard proof obligations. LTL consist of path formulae with the temporal operators X (next), F (future), G (global), U (until) and R (release). Expressions between curly braces are B predicates which can refer to the variables of the Event-B model. The following LTL formulae express the the state transitions and can be validated using the ProB model checker for ProB (also described in [Plagge and Leuschel, 2010]):

$$G(\{tl\_cars = \{green\}\} \implies$$
$$(\{tl\_cars = \{green\}\} \ U \{tl\_cars = \{yellow\}\})) \wedge$$
$$G(\{tl\_cars = \{yellow\}\} \implies$$
$$(\{tl\_cars = \{yellow\}\} \ U \{tl\_cars = \{red\}\})) \wedge$$
$$G(\{tl\_cars = \{red\}\} \implies$$
$$(\{tl\_cars = \{red\}\} \ U \{tl\_cars = \{red, yellow\}\})) \wedge$$
$$G(\{tl\_cars = \{red, yellow\}\} \implies$$
$$(\{tl\_cars = \{red, yellow\}\} \ U \{tl\_cars = \{green\}\}))$$

This formula is an equivalence to D3.2 and can be validated against the model using ProB.

The usefulness of this, particularly in this example, is questionable: A one-to-one equivalence has been established at the expense of a formula that is hard to read, and probably incomprehensible to the stakeholders. Still it demonstrates how other formalisms can be employed, as has been argued in Section 3.1.4.

An approach to validation that may be more accommodating to stakeholders is animation, as described in the next section.

5.5.2   Validation with Animation

As mentioned in Section 5.4.1, a weakness of Event-B is the representation of time and sequences. One could prove that none of the invariants of the model were violated — but this does not mean that the system behaves as expected. An extreme scenario would be a system that does not allow any state transitions. In such a system, none of the invariants could

be violated at all (assuming a valid initialisation), but that is rarely the
expected behaviour.

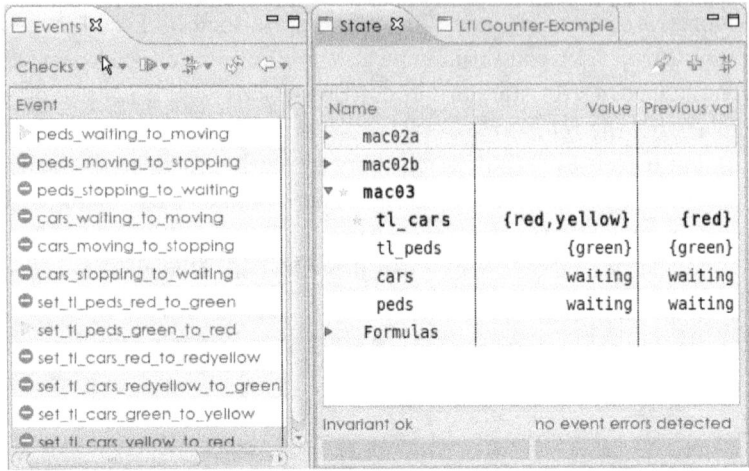

Figure 5.7: A valid but incorrect state in iteration 3. This is a screenshot
of ProB.

To validate a model with a domain expert, an animator could be
used. In this example, ProB has been used [Leuschel and Butler, 2003].
As Event-B relies on proofs, the model checking capabilities of ProB
are not needed here — although the model checker is a useful tool for
finding out why a proof cannot be discharged, for instance. As all proofs
were discharged in this model, the domain expert will not find invariant
violations. Instead, the objective would be to find incorrect behaviour.

And indeed, the model does not behave exactly as one would expect.
Figure 5.7 shows the model in ProB after a number of state transitions.
The pedestrians are waiting, but their light is green. Still, the model
allowed tl_cars to be set to { red, yellow}, because this still means "stop"
according to Figure 5.6.

This can be fixed by adding another requirement:

R3.1   If [tl_peds] is [green], then [tl_cars] must be [red].

This requirement in turn can be translated into an invariant, which in
turn requires a guard to be added, as otherwise one proof obligation could
not be discharged:

INVARIANTS

r3.1 : $tl\_peds = \{green\} \Rightarrow tl\_cars = \{red\}$

EVENTS

Event $set\_tl\_cars\_red\_to\_redyellow \; \widehat{=}$

refines set_cars_signal

...

when

s3.3 : $tl\_peds \neq \{green\}$

...

5.5.3 Adding Implementation Detail

A domain expert may object the fact that the green cycles of the traffic lights for cars and pedestrians do not have to alternate in this model: [tl_peds] could cycle endlessly between [red] and [$green$], without [tl_cars] turning green at all. This is not surprising, as this requirement is not captured anywhere. In fact, it takes a domain expert to decide whether this is even true (there may be special situations like maintenance where such a requirement does not hold).

For this case study, this requirement is now recorded explicitly:

R3.2	[$green$] cycles for [tl_cars] and [tl_peds] must alternate.

It is tricky to capture this requirement as an invariant because of its temporal nature. The ProR approach provides several ways of dealing with this. One option is to simply not model it formally. To still demonstrate consistency of the model, the corresponding specification element could be state informally, thereby satisfying (3.4). Another option would by using a formalism that allows expressing R3.2, like LTL, as demonstrated in Section 5.5.1.

Instead, this requirement could be implemented using an auxiliary variable. But such a variable is hidden from the environment, an would therefore be part of s_h, which, according to Section 3.2.1, makes it implementation P. In Section 3.2.2 it was argued that, while implementation is not the focus of the ProR approach, it can still be employed if deemed useful. Further, there is no difference in principal on whether refinement is employed for specification or implementation, as discussed in Section 3.3.2.

To demonstrate this, the realisation of R3.2 with implementation is demonstrated in the following. For clarity, this is realised in a separate refinement mac03b. The following machine has been shortened to only show the relevant model elements, the full machine is Available in Appendix B.4:

MACHINE mac03b
REFINES mac03a
SEES ctx03
VARIABLES

    peds_was_green
EVENTS
Event *set_tl_peds_red_to_green* $\widehat{=}$
extends *set_tl_peds_red_to_green*
  when

      p3.1a : $peds\_was\_green = FALSE$
  end
Event *set_tl_peds_green_to_red* $\widehat{=}$
extends *set_tl_peds_green_to_red*
  then

      p3.1b : $peds\_was\_green := TRUE$
  end
Event *set_tl_cars_red_to_redyellow* $\widehat{=}$
extends *set_tl_cars_red_to_redyellow*
  when

      p3.1c : $peds\_was\_green = TRUE$
  end
Event *set_tl_cars_yellow_to_red* $\widehat{=}$
extends *set_tl_cars_yellow_to_red*
  then

      p3.1d : $peds\_was\_green := FALSE$
  end
END

This implementation results in the following traceability:

R3.2 ↔ p3.1a, p3.1b, p3.1c, p3.1d

It may be useful to annotate this trace, which is supported by the tool
ProR.

5.6   Iteration 4: Modelling the Buttons

Let's continue with the process shown in Figure 3.4 and formalise the next
requirement R1.2:

R1.2	[*peds*] signal their wish to cross the [*street*] by [*push*]ing one of the [*button*]s.

Refinement is applied here to gradually include formal requirements into subsequent refinements, as described in Section 3.3. This can be demonstrated by adding a variable [*request*] for the managing the push button state, as visualised in Figure 5.2 by the phenomenon PB!{pushed, not pushed}, as well as the TC!{reset}.

The state of the variable is controlled by both the environment (e_v, a pedestrian triggering the request to cross) and the system (by resetting the request to cross). The distinction between the two is not made with respect to the variable itself, but with respect to the events modifying the [*request*] state.

Pedestrians can push the button anytime, as often as they want. However, the request can only be set if it is not set yet. This distinction is reflected by Figure 5.2. In the formal model, the pedestrian and the pushing of the button is not modelled, only the setting of the request flag. This results in additional requirements:

R4.1	[*push*]ing a [*button*] results in [*request*] to be set, it not yet set.
R4.2	If [*request*] is set and [*tl_peds*] is not [*green*], it will eventually turn [*green*].
R4.3	After [*tl_peds*] turns [*red*], [*request*] is reset.

Further, R1.2 contains some non-functional aspects and will therefore evolve in a non-functional requirement. Doing so will ensure that it still has to be justified, but also that R4.1, R4.2 and R4.3 can participate in the justification, as expressed by (3.7):

N4.4	[*peds*] signal their wish to cross the [*street*] by [*push*]ing one of the [*button*]s.

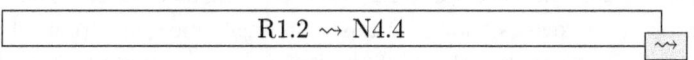

R1.2 ⤳ N4.4

This is now sufficient information for modelling the new requirements in refinement mac04. To model the setting of [*request*] by the environment, a new event set_request is introduced. The resetting is realised by adding actions to the existing events set_tl_peds_green_to_red and set_tl_cars_green_to_yellow, which are controlled by the system.

MACHINE mac04

REFINES   mac03

SEES   ctx03

VARIABLES

    request

INVARIANTS

       type_request : $request \in BOOL$

EVENTS

Event   *set_ request* $\widehat{=}$

    when

          s4.1a : $request = FALSE$

    begin

          s4.1b : $request := TRUE$

    end

Event   *set_ tl_ cars_ green_ to_ yellow* $\widehat{=}$

extends *set_ tl_ cars_ green_ to_ yellow*

    when

          s4.2 : $request = TRUE$

    end

Event   *set_ tl_ peds_ green_ to_ red* $\widehat{=}$

extends *set_ tl_ peds_ green_ to_ red*

    then

          s4.3 : $request := FALSE$

    end

END

R4.1 and R4.3 can be traced directly into the model:

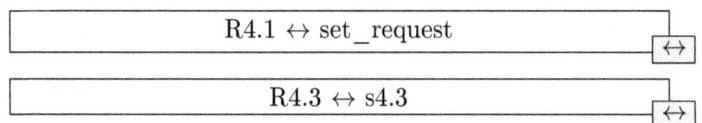

This is not the case for R4.2, however. In fact, closer inspection of R4.2 and r4.2 reveals that r4.2 has a stronger meaning than what R4.2 expresses: It states that the traffic light for cars may only turn red, if, and only if, the signal has been requested. This may be stronger than what is needed. For instance, there may be a maintenance mode that allows turning the signal to red. For simplicity in this case study, this meaning is retained, but made explicit.

Further, R4.2 contains a temporal statement that is not part of the formal model. Therefore, R4.2 is rephrased, and split into two:

R4.2′	[tl_cars] may only turn not [green], it a [request] is pending.
R4.2″	If a [request] is pending, [tl_peds] must eventually turn [green].

The second requirement R4.2″ contains a temporal property. A simple option is the conversion into a non-functional property. Another alternative would be the creation of a temporal property, corresponding to Section 5.5.1.

5.7 Iteration 5: Introducing Time

As discussed in Section 5.4.1, time can be tricky with a formalism like Event-B. The requirements contain two requirements that introduce timing constrains, which are N1.2 and N1.3. These were modelled as non-functional requirements.

It is possible to evolve these requirements in a functional and non-functional component. Specifically, it is possible to model an abstract clock in Event-B that counts "ticks", where each tick represents a certain amount of time (the length of a tick is specified as an informal artefact). Such ticks are introduced for t_1 (N1.2) and t_2 (N1.3).

While designing in Event-B, the question of the duration of each state transition (event) arises. This can be recorded as a non-functional requirement. and has to be validated outside the formal model, according to (3.7). With this idea, N1.2 and N1.3 evolve as follows:

N5.1	Each event that modifies traffic light transitions is [1] [tick] long.
N5.2	The length of a [tick] is [1 second] with a tolerance of 5%.
R5.1	Between a [request] and [tl_peds] turing [green], at most [t_1] ticks must pass.
R5.2	Upon turning [green], [tl_peds] must stay green for [t_2] ticks.

Note that in practice, there would be several other timing constants, i.e. the minimum amount of time [tl_cars] needs to stay [red], after [tl_peds] turned red. To keep this example reasonably short, only the two timing properties above will be modelled.

Now the constants t_1 and t_2 can be introduced in a new context ctx05. It can be useful to temporarily assign concrete numbers to such constants to ease model checking.

Reasoning over the model and these artefacts unveils that R5.1 may allow values for $[t\_1]$ that are not feasible. Specifically, there is a lower limit. By reasoning over the model, this value can be found: A pedestrian can request crossing as soon as $[tl\_peds]$ turns red. For $[tl\_cars]$ to turn $[green]$ and then $[red]$ again, and then $[tl\_peds]$ to turn $[green]$, at least 5 $[tick]$s will pass. While this has just been demonstrated with reasoning, a model check should be able to confirm this on the completed model.

But from that point of view, R5.1 makes only sense if there is a reason to not react to a crossing request right away. And there is, to give the cars a chance to cross the intersection as well. Thus, a missing requirement has been identified:

R5.3   Upon turning $[green]$, $[tl\_cars]$ must stay green for $[t\_3]$ ticks.

Also, the state that tracks the ticks is, once again, implementation detail. Taken all these things into consideration, the machine can be constructed. A new refinement mac05 is created that introduces two variables for counting the ticks for the green phases. Upon turning green, the counters are set to the corresponding value. New "tick" events count backwards to ensure that the traffic lights stay $[green]$ for the minimum required time. The resulting machine can be found in Appendix B.6.

5.7.1   Hidden Domain Properties

The model so far still contains state and events that model hidden domain properties, $[peds]$ and $[cars]$. These can be useful, as they help to understand the model. At the same time, it should be possible to remove them form the model without affecting the desired functionality. Event-B does not allow this. It can be simulated, however, by simply adding guards that evaluate to false ($\bot$) to those events. Doing so is particularly helpful when inspecting the state space manually, as it is desirable to keep the state space as small as possible. This is shown in Figure 5.8, where a state diagram has been generated by ProB. As an additional measure to keep the state space small, the constants $[t\_1]$ and $[t\_2]$ have been set to 1.

As the state space of this model is rather small, it is well-suited for inspection and can be used to validate the model against the requirements.

Alternatively, some of the requirements could have been stated using LTL, as described in Section 5.5.1. Whether this would make things easier, depends on the needs of the project.

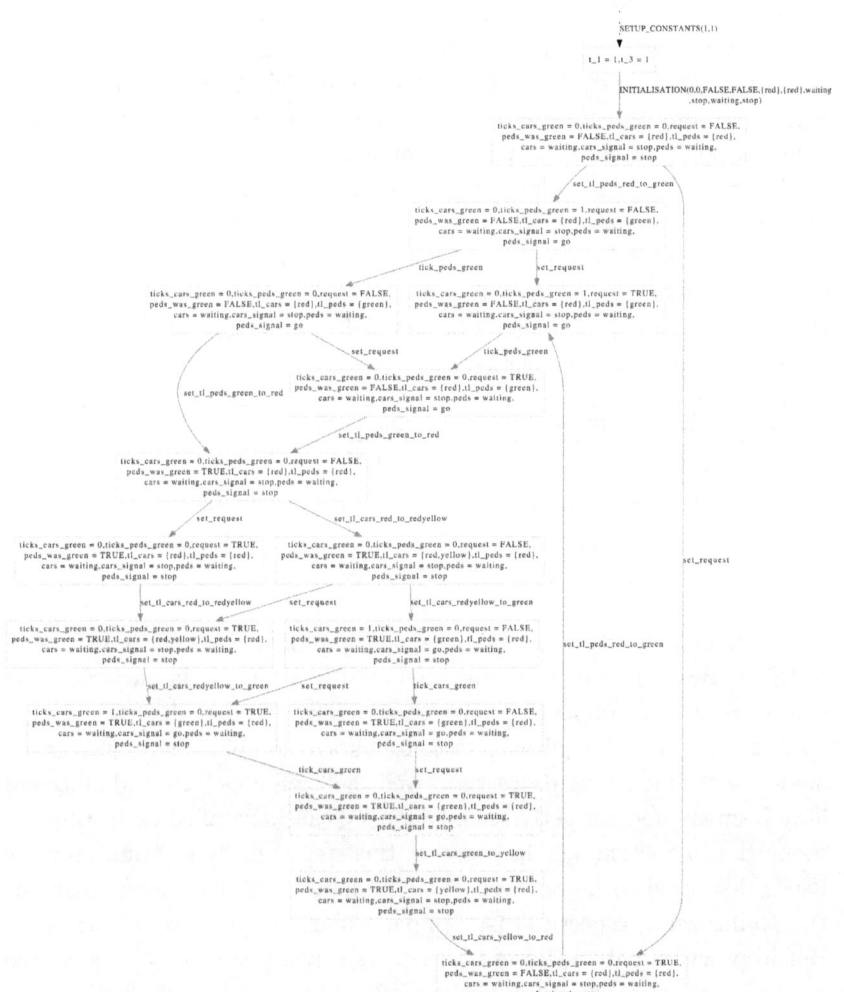

Figure 5.8: The state space created by ProB (all ticks reduced to 1, to keep the state space small)

5.8 Analysis and Conclusion

One of the reasons for building the model and creating the traces was to validate the model's consistency. For a consistent model, the properties presented in Section 3.6 must hold. The system description and formal model so far can be found in Appendix B.6. The informal system description consists of 11 functional requirements, 5 non-functional requirements,

2 design decisions, 1 specification artefact and 16 domain properties.

The fact that only one specification artefact exists is misleading, as many more exist, but in the formal model, without an informal representation. Whether an informal element needs to exist is up to the modeller, and not only for specification elements.

The model shown so far is not consistent, because R1.3 has not been realised yet (compliance with regulations, RiLSA). As RiLSA contains many requirements concerning the domain (i.e. arrangement of hardware), its realisation is not possible. Therefore, it has to be validated outside the model. In practice, R1.3 would be realised by passing certification. This could be documented as follows:

S6.1   The system passes the certification process

With this, the traceability for R1.3 could be completed:

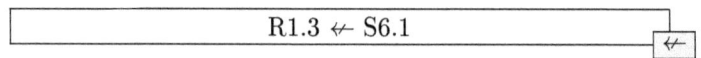

Unfortunately, this trace stays suspect until the system is built and certified. But this is now documented. Other artefacts, in particular the non-functional requirements, must be handled correspondingly.

A subset of the requirements contains a traceability into the formal model. If the modelling process followed the **ProR approach**, and all traces have been created correctly, then adequacy can be verified by testing the properties in Section 3.6. In practice, this step is likely to bring up more issues that need to be resolved. Crucial here is the completeness of the traceability with respect to (3.4). In particular, (3.4) involves all artefacts. But in practice, only a subset of artefacts called satisfaction base is used for reasoning, as discussed in Section 3.2.3. If the chosen satisfaction base is too small (meaning that there are traces missing), then the argument may be flawed. This is not a problem for the formal model, as by proving, a correct satisfaction base is selected (otherwise, the proof obligation could not be discharged).

The uses trace can be verified as well with respect to the categorisation of phenomena and their association with artefacts, as captured in (3.18) – (3.28).

The uses relationship could be used even further, in principle. For instance, in equivalence relationships it could be demanded that both the formal and informal artefacts use the same phenomena, i.e. $A^I \leftrightarrow A^F \Rightarrow uses[A^I] = uses[A^F]$. But this does not work, because the **ProR approach** does not distinguish between state (e.g. variables) and values (e.g. constants). Using a more sophisticated model for phenomena could remedy this and is a topic for future work.

The model was realised in Rodin, using the ProR plug-in, which was described in Section 4.7. Using the plug-in helped keeping the model consistent (with respect to the relations defined in Section 3.3. The colour highlighting of phenomena helped to quickly and reliably identify all phenomena. Once justification traces were established, the tool helped to systematically re-validate the traces that were pointing to artefacts that had changed (suspect links).

Working through the case study also helped identify weaknesses of the tool and sparked ideas on how to improve it. This has been discussed in Section 4.8.

Chapter 6

Conclusions and Future Work

This work is concerned with the **ProR approach**, an approach for incrementally building "high quality" system descriptions. The resulting system description consists of formal and informal artefacts, and is complemented by a traceability that is also built during the incremental development process. The resulting traceability supports systematic validation and change management.

This work is also concerned with the development of the ProR tool, an Open Source, Eclipse-based platform for working with requirements. ProR is based on an open standard and can be integrated with formal modelling tools to support the **ProR approach**.

6.1 Contributions

These are the two principal contributions of this work, the **ProR approach** and the ProR tool. This work is further complemented by a case study that demonstrates how the **ProR approach** is applied in practice, using ProR.

6.1.1 The ProR Approach

A central contribution of the **ProR approach** is the validation of the *informal* system description, by means of a formal model. While there has been significant progress in the field of formal model, a lot of research is concerned with the consistency of the formal model itself, leaving out its original purpose. This can result in formal specifications that are proven correct, but still don't achieve what the stakeholder intend.

A consequence of the **ProR approach** is the fact that, even though a formal model exists, the conclusions drawn are informal. This is due to (3.10), where premise $W^I \wedge S^I$ and conclusion $R^I \wedge D^I$ are both informal. Both are connected by the formal model by means of traceability, according to (3.11) and (3.12). Considering that the association is realised by informal means, the question arises what the actual value of the formal model (and formal proof is). But this question is not limited to the **ProR approach**, but applies to all methods that attempt to solve real-world problems with formal modelling. The temptation exists to identify the formal artefacts as requirements or domain properties. And this is indeed desirable (equivalence), but not always achievable.

From that point of few, the real value of the **ProR approach** stems from the traceability. The traceability supports informal reasoning, and acknowledges both that some traces are not equivalences (but justifications and realisations), and that some artefacts simply cannot be modelled formally at all. Further, the traceability provides robustness that provides confidence when changes are necessary. At the same time, it allows (but does not demand) to give formal modelling a central role in the system description.

6.1.2   The ProR Platform

With ProR, this work makes a practical contribution to the tool landscape in requirements engineering. Open source tools for managing requirements existed before, but due to a missing common standard, those tools were confined to niches and did not provide much interoperability. The emerging ReqIF standard provided such a missing standard, and ProR is the first Open Source implementation supporting this standard. By using Eclipse as the tool platform, integration with other tools is facilitated and encouraged. That an integration is possible and can be seamless has been demonstrated by integrating ProR with Rodin, a tool for Event-B modelling.

ProR has already been deployed in other academic projects, as discussed in Section 4.8. ProR also gained some visibility in industry, as was described in Section 4.2.2. As an Eclipse Foundation project, there is a good chance that ProR will survive this work and flourish both in academic and industrial use.

6.1.3 Case Study

To demonstrate the feasibility of the ProR approach and the usability of ProR in principle, a system description for a traffic light controller has been created. It showed that the approach works in principle. While the case study was not a real-world example, it was still large enough to demonstrate the various aspects of the ProR approach. It showed how artefacts were structured and how a formal model would be build iteratively, improving and extending the informal artefacts at the same time.

6.2 Future Work

This work represents a self-contained approach with tool support. While its feasibility has been demonstrated with a small example, a major goal is the application on a real-world project.

This work sparked a number of ideas on how the ProR approach could be extended. Particularly interesting is the further structuring of phenomena. The ProR approach takes a rather broad approach to classifying phenomena, by not distinguishing state, constants or events. Other approaches and notations, like Problem Frames, KAOS or UML, use more elaborate models for structuring phenomena. Using such approaches would result in additional properties for a consistent system description.

Another area of research concerns domain-specific languages (DSLs), which could make the tracing between informal and formal artefacts easier. In fact, depending on the DSL, the formal artefact could be generated automatically, although potentially at the expense of readability to the stakeholders. Some work on this has already been published in [Jastram and Graf, 2011d].

The use of animation has been discussed in Section 5.5.2. This could be taken one step further, by extracting more information from the system description to aid in the animation process. For instance, it should be possible to automatically generate Problem Frames diagrams and to animate them.

The work on the integration of ProR and Rodin will be continued. Specifically, the tool currently only supports the establishing of the traceability, but support for analysis is limited. It is desirable that the tool reports all violations of known consistency properties, as described in Section 3.1.3.

As discussed in Section 4.8, work on ProR as a generic tool for requirements engineering will continue as well.

6.3   Conclusions

The aim of this research was to "develop a practical approach for specifying systems that combines formal and informal specification methods to take advantage of their respective advantages and minimises their respective disadvantages" (Section 3.1). I believe that this research brings this goal significantly closer: The combination of formal and informal specification methods has been achieved by building a traceability theory and an approach for incrementally applying it to build a system description. The practical aspect has been realised by building the ProR tool, and by building a community and infrastructure that will ensure the survival of ProR beyond this work.

Appendix A

Eclipse Proposal for RMF

This is the latest version of the proposal of RMF to the Eclipse Foundation, before it became an official project in November 2011. The proposal can be found online at http://www.eclipse.org/proposals/modeling.mdt.rmf/.

The list of "interested parties" includes well-known companies like Airbus, Atos and itemis.

The Proposal

The "Requirements Modeling Framework" (RMF) project is a proposed open source project under the Model Development Tools Project.

This proposal is in the Project Proposal Phase (as defined in the Eclipse Development Process) and is written to declare its intent and scope. We solicit additional participation and input from the Eclipse community. Please send all feedback to the Eclipse Proposals Forum.

The vision is to have at least one clean-room implementation of the OMG ReqIF standard in form of an EMF model and some rudimentary tooling to edit these models. The idea is to implement the standard so that it is compatible with Eclipse technologies like GMF, Xpand, Acceleo, Sphinx, etc. and other key technologies like CDO.

Background

The Eclipse ecosystem provides a number of projects to support software development and systems engineering. However, in the open source community, one important aspect of the engineering process is very much neglected: requirements management, consisting of a number of sub-disciplines including requirements capturing, requirements engineering, requirements traceability, change management and product line engineering, to name just a few.

The goal of RMF is to provide the technical basis for open source projects that implement tools for requirements management. The conditions for the inception of such a project are perfect: Until now, all tools for requirements engineering suffered from the lack of a standard for the exchange and storage of requirements. Each tool provider invented his own method and meta-model for requirements, thereby limiting the common user base and the possibility for exchange between tools.

The OMG just released the Requirements Interchange Format (ReqIF), an XML-based data structure for exchanging requirements. The first draft of this standard with the name RIF was created in 2004, and various requirements tools (commercial and otherwise) already support it to some degree. Currently there are three actively used versions of the standard: RIF 1.1a, RIF 1.2 and ReqIF 1.0.1.

This open standard could have as much impact on requirements structuring as the UML had on modeling. The implementation of the ReqIF standard as an Eclipse project could similarly be as important for the requirements community as was the implementation of UML2 in Ecore for the modeling community by paving the way for such tools as Topcased and Papyrus MDT.

Providing such a project under the Eclipse umbrella would offer a possibility for many projects that are involved in requirements management to find a common implementation of the standard. It would push Eclipse in to phases of the development process where it is currently underrepresented.

Scope

The RMF project's focus is the creation of libraries and tools for working with ReqIF-based requirements. The objective is to provide the community with a solid implementation of the standard upon which various tools can be built. RMF will provide a means for data exchange between tools, an EMF-based data model, infrastructure tooling and a user interface.

RMF will not provide support for Requirements Management. Instead, it is expected that users will use specialized tools or work with the available Eclipse tooling (EMF Compare, version control integration, etc.). Generic or specific parts of the tooling can be hosted as part of the RMF project.

Description

The following diagram depicts the architecture of the current development and indicates which elements will be part of the initial contribution:

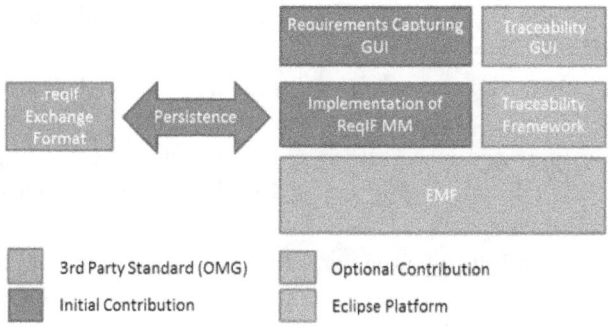

We created an EMF-based implementation of the ReqIF core that supports persistence using the RIF XML schema. Further, we created a GUI for capturing requirements.

These contributions have their origins in research projects, where they are actively used. In particular, these research projects already produced extensions, demonstrating the value of the platform.

Why Eclipse?

The Eclipse ecosystem will benefit from an implementation of a requirements standard to cover more aspects of system development. Currently, modelling is covered well on the low level (EMF, TMF, CDT, etc.) and high level (UML, SysML, etc.). Adding the domain of requirements capturing would extend the coverage. To promote it, we need not only a standard, but also a common implementation that tools build upon.

Being an Eclipse project will draw the interest of more parties to the project. The implementation of a standard benefits greatly from the participation of many parties (improvement of quality, reduction of cost). In addition, long-term support through the participation of many parties is essential for many domains.

Initial Contribution

The initial contribution will consist of an EMF-based ReqIF Core that supports persistence, and a tool front-end for working with ReqIF-data.

ReqIF Core

The first initial contribution will come from itemis and will include our implementation of the ReqIF and RIF metamodels as .ecore models, including special (de)serializers that map the EMF-models to a ReqIF conforming standard. The model and (de)serializers are already available at itemis and need only be provided.

The initial contribution has been internally tested. A ReqIF export from production data from the automotive domain from a commercial tool has been imported into Eclipse and exported again, keeping all the structural data, with the exception of ReqIFs XHTML extensions (see table below).

They have been implemented according to the specification, but since ReqIF is a new standard, no extensive tests with ReqIF files coming from other sources / other tools have been made.

RIF/ReqIF Version	EMF Metamodel	ProR Integration	XHTML Support	Tool extension support	Current industry support	Future relevance
RIF 1.1a	Available	Not available	Available	Not available	High	?
RIF 1.2	Available	Available	Available	Available	Low	?
ReqIF 1.0.1	Available	Not available	Available	Available	Low	High

GUI (known as ProR)

The second contribution will come from the University of Düsseldorf and will include a front end that facilitates working with ReqIF data (ProR). While ReqIF data could also be edited with the default EMF editor, this is not even remotely practical: a tree view of the requirements, with the details shown in the property view, doesn't allow users to efficiently navigate requirements or get an overview of what's there.

The GUI allows users to arrange only those requirements attributes that they care about in a grid view and implements a number of short-cuts for frequent operations that actually consist of a number of model transformations. Further, it contains an extension mechanism that allows integration with other EMF-based tools and supports custom rendering. The GUI currently only support RIF 1.2, and not all RIF features are implemented yet.

The GUI so far has been developed under the name ProR. This name, including the pror.org property, will be part of the contribution to the project (see legal issues below).

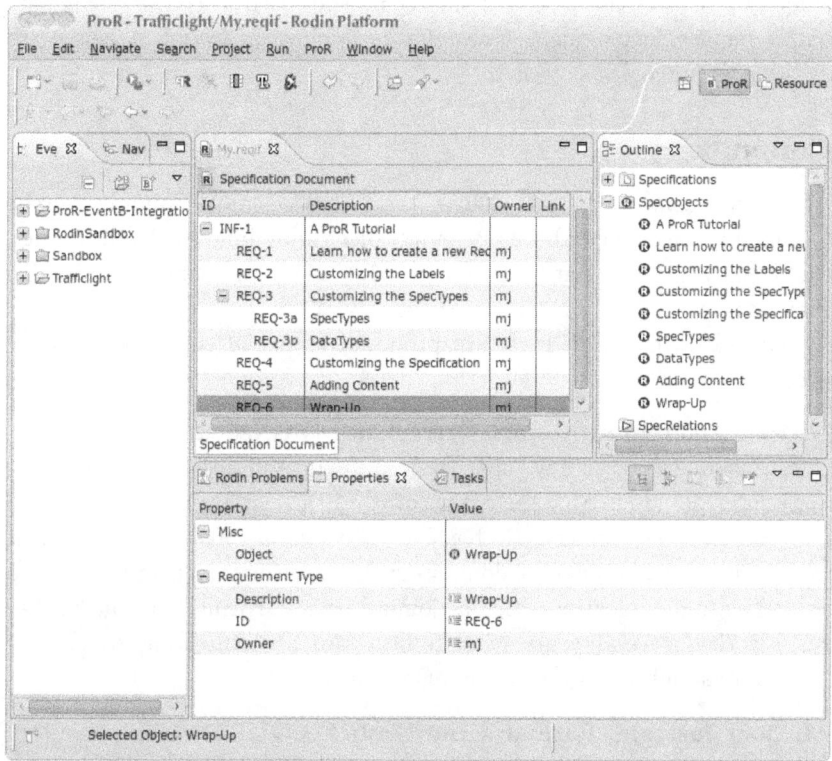

Legal Issues

All contributions will be distributed under the Eclipse Public License. The ReqIF metamodel has been fully developed by itemis. The GUI has been developed by University of Düsseldorf with changes from itemis.

The GUI development to date has been branded as ProR, supported by the pror.org website. The rights to the brand and the name reside with Michael Jastram, who is willing to agree to the Eclipse Trademark Transfer Agreement.

Related Projects

Sphinx: Requirements models tend to grow quite large in commercial projects. Using Sphinx will improve the performance and scalability. The current implementation is not yet based on / integrated with Sphinx.

EMF Compare: Since the RMF is based on EMF, EMF Compare could be a key technology for the comparison of requirements documents.

M2T/BIRT: With the new Indigo Release, BIRT includes an EMF adapter. BIRT could be one of the technlogies used to create documents out of requirement models. The M2T technologies (Xpand, Acceleo) are possible technologies as well.

Committers

The initial committers will deliver the initial release. Funding can be provided through the ITEA2 research project until June 2012 and through the Deploy project until February 2012. Funding may be available thereafter.

The following individuals are proposed as initial committers to the project:

Nirmal Sasidharan, itemis – RIF Core (Project Lead) Nirmal Sasidharan is a developer and software architect at itemis. His interests are in Model Driven Software Development (MDSD) based on Eclipse platform. He has over 10 years of software development experience in different domains such as Automotive, Aerospace and Telecommunication. Before joining itemis, Nirmal Sasidharan has worked several years with Robert Bosch architecting tools. He works and lives in Stuttgart, Germany.

Michael Jastram, Formal Mind GmbH – GUI Michael is coauthor of the German Book "Eclipse Rich Client Platform" and has been working with Java technologies as developer and architect since 1996. He is currently pursuing a Ph.D. in Computer Science at the University of Düsseldorf. He is founder and managing director of Formal Mind GmbH. He also founded and runs the local Java User Group (rheinjug). He holds a Master degree from M.I.T.

Lukas Ladenberger, University of Düsseldorf – GUI Lukas is currently a Ph.D. student in Computer Science at the University of Düsseldorf and an employee at Formal Mind GmbH. He has been working with Java and Eclipse as developer since 2004. He is also an active member of the local Java User Group (rheinjug).

Andreas Graf, itemis Andreas Graf is a Business Development Manager at the automotive division of itemis. He is an expert in MDSD for automotive software. Apart from his managerial role at itemis, Andreas is writing tools based on Eclipse platform. Before joining itemis, he has worked several years with BMW in the areas of process definition, ECU software development and Software logistics.

We welcome additional committers and contributions.

Mentors

The following Architecture Council member will mentor this project:

- Ed Merks
- Kenn Hussey

Interested Parties

The following individuals, organisations, companies and projects have expressed interest in this project:

- Airbus
- Atos
- emergn Ltd
- Formal Mind
- Heinrich-Heine University Düsseldorf
- HOOD GmbH
- itemis AG
- MKS
- ModelAlchemy Consulting
- Obeo
- Prostep
- TCL Software Ltd. (LuisCM)
- University of Applied Sciences Darmstadt (Prof. Fromm)

Project Scheduling

Initial contribution is anticipated in July or August 2011.

Changes to this Document

Date	Change
29-Jun-2011	Document created
25-Jul-2011	Added more interested parties; provided a link to the ProR website; fixed umlauts.
26-Jul-2011	Added even more interested parties; Updated the feature table (XHTML Support)
27-Jul-2011	Added one more committer biographies of committers. Updated the feature table (XHTML and Tool Support)
28-Jul-2011	Added information about the transfer of the ProR trademark to the Eclipse Foundation.

Appendix B

Case Study Model

B.1   Iteration 0

Iteration 0 represents the initial artefacts as produced by the stakeholders. No formal model exists for this iteration.

B.1.1   Artefacts

A0.1	The system allows pedestrians to cross the street safely
A0.2	The road is equipped with two traffic lights for the cars (colors red, yellow and green), one in each direction.
A0.3	The road is equipped with two traffic lights for the pedestrians (colors red and green), one on each side of the street.
A0.4	The traffic lights for the pedestrians are equipped with push buttons.
A0.5	The traffic for cars is usually green.
A0.6	Pedestrians can request permission to cross the street by pushing the push button.
A0.7	Pedestrians will get permission to cross the street t_1 seconds after the push button got pressed.
A0.8	The duration of the green light for pedestrians is t_2 seconds.
A0.9	The traffic light system follows the regulations for traffic lights of Germany (Richtlinien für Signalanlagen, RiLSA).

B.2   Iteration 1

Iteration 1 represents the artefacts that were restructured by using Problem Frames. No formal model exists for this iteration.

B.2.1   Artefacts

R1.1	The system allows [peds] [moving] across the [street] safely
R1.2	[peds] signal their wish to cross the [street] by [push]ing one of the [button]s.
R1.3	The traffic light system follows the regulations for traffic lights of Germany (Richtlinien für Signalanlagen, RiLSA)
N1.1	The traffic lights for cars [tl_cars] are usually [green].
N1.2	Between [push]ing the button for the first time in a cycle and [tl_peds] allowing pedestrians to cross, at most [t_1] seconds must pass.
N1.3	Upon turning [green], [tl_peds] must stay green for [t_2] seconds, with a tolerance of 5%.
W1.1	Two synchronised traffic lights for cars [tl_cars] are located on the [street], according to Figure 5.1
W1.2	The state of the car traffic lights is represented by [tl_cars], which represents a subset of [red], [yellow] and [green], meaning that the corresponding light is on.
W1.3	Two synchronised traffic lights for pedestrians [tl_peds] are located on the [street], according to Figure 5.1.
W1.4	The state of the pedestrian traffic lights is represented by [tl_peds], which represents a subset of [red] and [green], meaning that the corresponding light is on.
W1.5	Two buttons are mounted on the bases of the pedestrian traffic lights [tl_peds], according to Figure 5.1, allowing [peds] to [push] them.
W1.6	[peds] can press any of the push [button]s to trigger a [push] event.

B.3 Iteration 2

The second iteration is the first iteration where formal modelling takes place. The model consists of one context and two machines.

B.3.1 Artefacts

R2.1	When [peds] are [moving] or [stopping] on the [street], [cars] must be [waiting].
R1.2	[peds] signal their wish to cross the [street] by [push]ing one of the [button]s.
R1.3	The traffic light system follows the regulations for traffic lights of Germany (Richtlinien für Signalanlagen, RiLSA)
N2.1	The system has additional safety properties.
N1.1	The traffic lights for cars [tl_cars] are usually [green].
N1.2	Between [push]ing the button for the first time in a cycle and [tl_peds] allowing pedestrians to cross, at most [t_1] seconds must pass.
N1.3	Upon turning [green], [tl_peds] must stay green for [t_2] seconds, with a tolerance of 5%.
W1.1	Two synchronised traffic lights for cars [tl_cars] are located on the [street], according to Figure 5.1.
W1.2	The state of the car traffic lights is represented by [tl_cars], which represents a subset of [red], [yellow] and [green], meaning that the corresponding light is on.
W1.3	Two synchronised traffic lights for pedestrians [tl_peds] are located on the [street], according to Figure 5.1.
W1.4	The state of the pedestrian traffic lights is represented by [tl_peds], which represents a subset of [red] and [green], meaning that the corresponding light is on.
W1.5	Two buttons are mounted on the bases of the pedestrian traffic lights [tl_peds], according to Figure 5.1, allowing [peds] to [push] them.
W1.6	[peds] can press any of the push [button]s to trigger a [push] event.
W2.1	[peds] that are not on the [street] are [waiting]. Upon entering the [street], they are [moving], followed by [stopping], before [waiting] again.
W2.2	[cars] that are not on the [street] are [waiting]. Upon entering the [street], they are [moving], followed by [stopping], before [waiting] again.

W2.3   Upon activating the system, [peds] are [waiting].

W2.4   Upon activating the system, [cars] are [waiting].

W2.5   Conceptually, the traffic lights [peds_ signal] and [cars_ signal] can indicate a [stop] or [go] signal, which is represented in the form of colours.

W2.6   The initial state for [peds_ signal] and [cars_ signal] is [stop]

W2.7   [peds] start [moving] only if [peds_ signal] is [go].   If [peds_ signal] turns to [stop], [peds] that are [moving] are [stopping] and will be [waiting], once they finished crossing.

W2.8   [cars] start [moving] only if [cars_ signal] is [go].   If [cars_ signal] turns to [stop], [cars] that are [moving] are [stopping] and will be [waiting], once they finished crossing.

W2.9   When [peds_ signal] indicates [go], the [cars] are [waiting].

W2.10   When [cars_ signal] indicates [go], the [peds] are [waiting].

S2.1   [tl_peds] and [tl_ cars] must never be [go] at the same time.

B.3.2   Context ctx02

CONTEXT   ctx02

SETS

   $MOVING$

   $SIGNAL$

CONSTANTS

   $moving$

   $stopping$

   $waiting$

   $stop$

   go

AXIOMS

   axm1 : $partition(MOVING, \{moving\}, \{stopping\}, \{waiting\})$

   axm2 : $partition(SIGNAL, \{stop\}, \{go\})$

END

B.3.3   Machine mac02a

MACHINE   mac02a

SEES   ctx02

VARIABLES

    peds
    cars

    peds_signal
    cars_signal

INVARIANTS

    w2.1 : $peds \in MOVING$
    w2.2 : $cars \in MOVING$
    w2.5a : $peds\_signal \in SIGNAL$
    w2.5b : $cars\_signal \in SIGNAL$

EVENTS

Initialisation

    begin

        w2.3 : $peds := waiting$
        w2.4 : $cars := waiting$
        w2.6a : $peds\_signal := stop$
        w2.6b : $cars\_signal := stop$

    end

Event *peds_waiting_to_moving* $\widehat{=}$

    when

        w2.1a : $peds = waiting$
        w2.7 : $peds\_signal = go$

    then

        w2.1b : $peds := moving$

    end

Event *peds_moving_to_stopping* $\widehat{=}$

    when

        w2.1c : $peds = moving$

    then

        w2.1d : $peds := stopping$

    end

Event *peds_stopping_to_waiting* $\widehat{=}$

    when

        w2.1e : $peds = stopping$

    then

        w2.1f : $peds := waiting$

    end

Event   *cars_waiting_to_moving* $\widehat{=}$
   when

        w2.2a : *cars = waiting*
        w2.8 : *cars_signal = go*
   then

        w2.2b : *cars := moving*
   end

Event   *cars_moving_to_stopping* $\widehat{=}$
   when

        w2.2c : *cars = moving*
   then

        w2.2d : *cars := stopping*
   end

Event   *cars_stopping_to_waiting* $\widehat{=}$
   when

        w2.2e : *cars = stopping*
   then

        w2.2f : *cars := waiting*
   end

Event   *set_peds_signal* $\widehat{=}$
   any

        signal
   where

        w2.5c : $signal \in SIGNAL$
   then

        w2.5d : *peds_signal := signal*
   end

Event   *set_cars_signal* $\widehat{=}$
   any

        signal
   where

        w2.5e : $signal \in SIGNAL$
   then

        w2.5f : *cars_signal := signal*
   end
END

B.3.4 Machine mac02b

MACHINE mac02b

REFINES mac02a

SEES ctx02

VARIABLES

> *peds*
> *cars*
>
> *peds_signal*
> *cars_signal*

INVARIANTS

> r2.1 : $(peds = moving \lor peds = stopping) \Rightarrow cars = waiting$
>
> w2.9 : $peds\_signal = go \Rightarrow cars = waiting$
>
> w2.10 : $cars\_signal = go \Rightarrow peds = waiting$
>
> s2.1 : $\neg(peds\_signal = go \land cars\_signal = go)$

EVENTS

Initialisation

> *extended*
>
> **begin**
>
> > w2.3 : $peds := waiting$
> > w2.4 : $cars := waiting$
> > w2.6a : $peds\_signal := stop$
> > w2.6b : $cars\_signal := stop$
>
> **end**

Event *peds_waiting_to_moving* $\widehat{=}$
extends *peds_waiting_to_moving*

> **when**
>
> > w2.1a : $peds = waiting$
> > w2.7 : $peds\_signal = go$
>
> **then**
>
> > w2.1b : $peds := moving$
>
> **end**

Event *peds_moving_to_stopping* $\widehat{=}$
extends *peds_moving_to_stopping*

> **when**
>
> > w2.1c : $peds = moving$

then

    w2.1d : $peds := stopping$

 end

Event   *peds_stopping_to_waiting* $\;\widehat{=}$

extends *peds_stopping_to_waiting*

 when

    w2.1e : $peds = stopping$

 then

    w2.1f : $peds := waiting$

 end

Event   *cars_waiting_to_moving* $\;\widehat{=}$

extends *cars_waiting_to_moving*

 when

     w2.2a : $cars = waiting$

     w2.8 : $cars\_signal = go$

 then

    w2.2b : $cars := moving$

 end

Event   *cars_moving_to_stopping* $\;\widehat{=}$

extends *cars_moving_to_stopping*

 when

     w2.2c : $cars = moving$

 then

    w2.2d : $cars := stopping$

 end

Event   *cars_stopping_to_waiting* $\;\widehat{=}$

extends *cars_stopping_to_waiting*

 when

     w2.2e : $cars = stopping$

 then

    w2.2f : $cars := waiting$

 end

Event   *set_peds_signal* $\;\widehat{=}$

extends *set_peds_signal*

 any

    signal

where

> w2.5c : $signal \in SIGNAL$
> grd1 : $cars = waiting$
> s2.1a : $\neg(signal = go \land cars\_signal = go)$

then

> w2.5d : $peds\_signal := signal$

end

Event $set\_cars\_signal \,\widehat{=}$
extends set_cars_signal

any

> $signal$

where

> w2.5e : $signal \in SIGNAL$
> grd2 : $peds = waiting$
> s2.1b : $\neg(peds\_signal = go \land signal = go)$

then

> w2.5f : $cars\_signal := signal$

end

END

B.4   Iteration 3

In the third iteration, the traffic light signals will be mapped from stop
and go to the actual colours, using data refinement.

B.4.1   Artefacts

R2.1	When [peds] are [moving] or [stopping] on the [street], [cars] must be [waiting].
R1.2	[peds] signal their wish to cross the [street] by [push]ing one of the [button]s.
R1.3	The traffic light system follows the regulations for traffic lights of Germany (Richtlinien für Signalanlagen, RiLSA)
R3.1	If [tl_peds] is [green], then [tl_cars] must be [red].
R3.2	[green] cycles for [tl_cars] and [tl_peds] must alternate.
N2.1	The system has additional safety properties.
N1.1	The traffic lights for cars [tl_cars] are usually [green].
N1.2	Between [push]ing the button for the first time in a cycle and [tl_peds] allowing pedestrians to cross, at most [t_1] seconds must pass.
N1.3	Upon turning [green], [tl_peds] must stay green for [t_2] seconds, with a tolerance of 5%.
D3.1	The relationship between [tl_peds] and [peds_signal] is a mapping of [red] to [stop] and [green] to [go], respectively.
D3.2	The relationship between [tl_cars] and [cars_signal] shall adhere to Figure 5.6.
W1.1	Two synchronised traffic lights for cars [tl_cars] are located on the [street], according to Figure 5.1.
W1.2	The state of the car traffic lights is represented by [tl_cars], which represents a subset of [red], [yellow] and [green], meaning that the corresponding light is on.
W1.3	Two synchronised traffic lights for pedestrians [tl_peds] are located on the [street], according to Figure 5.1.
W1.4	The state of the pedestrian traffic lights is represented by [tl_peds], which represents a subset of [red] and [green], meaning that the corresponding light is on.
W1.5	Two buttons are mounted on the bases of the pedestrian traffic lights [tl_peds], according to Figure 5.1, allowing [peds] to [push] them.
W1.6	[peds] can press any of the push [button]s to trigger a [push] event.

W2.1	[peds] that are not on the [street] are [waiting]. Upon entering the [street], they are [moving], followed by [stopping], before [waiting] again.
W2.2	[cars] that are not on the [street] are [waiting]. Upon entering the [street], they are [moving], followed by [stopping], before [waiting] again.
W2.3	Upon activating the system, [peds] are [waiting].
W2.4	Upon activating the system, [cars] are [waiting].
W2.5	Conceptually, the traffic lights [peds_ signal] and [cars_ signal] can indicate a [stop] or [go] signal, which is represented in the form of colours.
W2.6	The initial state for [peds_ signal] and [cars_ signal] is [stop]
W2.7	[peds] start [moving] only if [peds_ signal] is [go]. If [peds_ signal] turns to [stop], [peds] that are [moving] are [stopping] and will be [waiting], once they finished crossing.
W2.8	[cars] start [moving] only if [cars_ signal] is [go]. If [cars_ signal] turns to [stop], [cars] that are [moving] are [stopping] and will be [waiting], once they finished crossing.
W2.9	When [peds_ signal] indicates [go], the [cars] are [waiting].
W2.10	When [cars_ signal] indicates [go], the [peds] are [waiting].
S2.1	[tl_peds] and [tl_cars] must never be [go] at the same time.

B.4.2   Context ctx03

CONTEXT   ctx03
EXTENDS   ctx02
SETS

    COLOURS
CONSTANTS

    red

    yellow

    green
AXIOMS

    colours $: partition(COLOURS, \{red\}, \{yellow\}, \{green\})$
END

B.4.3   Machine mac03a

MACHINE   mac03a

REFINES   mac02b

SEES   ctx03

VARIABLES

    peds

    cars

    tl_peds

    tl_cars

INVARIANTS

    d3.1a : $tl\_peds \in \{\{red\}, \{green\}\}$

    d3.2a : $tl\_cars \in \{\{red\}, \{yellow\}, \{green\}, \{red, yellow\}\}$

    d3.1b : $peds\_signal = go \Leftrightarrow green \in tl\_peds$

    d3.2b : $cars\_signal = go \Leftrightarrow green \in tl\_cars$

    r3.1 : $tl\_peds = \{green\} \Rightarrow tl\_cars = \{red\}$

EVENTS

Initialisation

    begin

        init3.0 : $peds := waiting$

        init3.1 : $cars := waiting$

        init3.2 : $tl\_peds := \{red\}$

        init3.3 : $tl\_cars := \{red\}$

    end

Event   *peds_waiting_to_moving* $\widehat{=}$

refines *peds_waiting_to_moving*

    when

        w2.1a : $peds = waiting$

        s3.1c : $tl\_peds = \{green\}$

    then

        w2.1b : $peds := moving$

    end

Event   *peds_moving_to_stopping* $\widehat{=}$

extends *peds_moving_to_stopping*

    when

        w2.1c : $peds = moving$

    then

        w2.1d : $peds := stopping$

end

Event *peds_ stopping_ to_ waiting* $\widehat{=}$
extends *peds_ stopping_ to_ waiting*
   when

        w2.1e : $peds = stopping$
   then

        w2.1f : $peds := waiting$
   end

Event *cars_ waiting_ to_ moving* $\widehat{=}$
refines *cars_ waiting_ to_ moving*
   when

        w2.2a : $cars = waiting$
        s3.2c : $tl\_cars = \{green\}$
   then

        w2.2b : $cars := moving$
   end

Event *cars_ moving_ to_ stopping* $\widehat{=}$
extends *cars_ moving_ to_ stopping*
   when

        w2.2c : $cars = moving$
   then

        w2.2d : $cars := stopping$
   end

Event *cars_ stopping_ to_ waiting* $\widehat{=}$
extends *cars_ stopping_ to_ waiting*
   when

        w2.2e : $cars = stopping$
   then

        w2.2f : $cars := waiting$
   end

Event *set_ tl_peds_ red_ to_ green* $\widehat{=}$
refines *set_ peds_ signal*
   when

        s3.1d : $tl\_peds = \{red\}$
        s3.1e : $tl\_cars = \{red\}$
        grd1 : $cars = waiting$

    with

        signal : $signal = go$
    then

        s3.1f : $tl\_peds := \{green\}$
    end

Event *set_ tl_ peds_ green_ to_ red* $\widehat{=}$
refines *set_ peds_ signal*

    when

        s3.1g : $tl\_peds = \{green\}$
        grd1 : $cars = waiting$
    with

        signal : $signal = stop$
    then

        s3.1h : $tl\_peds := \{red\}$
    end

Event *set_ tl_ cars_ red_ to_ redyellow* $\widehat{=}$
refines *set_ cars_ signal*

    when

        grd2 : $peds = waiting$
        s3.2d : $tl\_cars = \{red\}$
        s3.3 : $tl\_peds \neq \{green\}$
    with

        signal : $signal = stop$
    then

        s3.2f : $tl\_cars := \{red, yellow\}$
    end

Event *set_ tl_ cars_ redyellow_ to_ green* $\widehat{=}$
refines *set_ cars_ signal*

    when

        grd2 : $peds = waiting$
        s3.2g : $tl\_cars = \{red, yellow\}$
        s3.2h : $tl\_peds \neq \{green\}$
    with

        signal : $signal = go$
    then

        s3.2i : $tl\_cars := \{green\}$

end

Event *set_tl_cars_green_to_yellow* $\widehat{=}$

refines *set_cars_signal*

 when

   grd2 : *peds = waiting*
   s3.2j : *tl_cars = {green}*
  with

   signal : *signal = stop*
  then

   s3.2k : *tl_cars := {yellow}*
  end

Event *set_tl_cars_yellow_to_red* $\widehat{=}$

refines *set_cars_signal*

 when

   grd2 : *peds = waiting*
   s3.21 : *tl_cars = {yellow}*
  with

   signal : *signal = stop*
  then

   s3.2m : *tl_cars := {red}*
  end

END

B.4.4   Machine mac03b

MACHINE mac03b

REFINES mac03a

SEES ctx03

VARIABLES

 peds
 cars

 tl_peds
 tl_cars
 peds_was_green

INVARIANTS

  type : *peds_was_green* $\in$ *BOOL*

EVENTS
Initialisation
 extended
 begin

   init3.0 : $peds := waiting$
   init3.1 : $cars := waiting$
   init3.2 : $tl\_peds := \{red\}$
   init3.3 : $tl\_cars := \{red\}$
   init : $peds\_was\_green := FALSE$
 end
Event *peds_waiting_to_moving* $\widehat{=}$
extends *peds_waiting_to_moving*
 when

   w2.1a : $peds = waiting$
   s3.1c : $tl\_peds = \{green\}$
 then

   w2.1b : $peds := moving$
 end
Event *peds_moving_to_stopping* $\widehat{=}$
extends *peds_moving_to_stopping*
 when

   w2.1c : $peds = moving$
 then

   w2.1d : $peds := stopping$
 end
Event *peds_stopping_to_waiting* $\widehat{=}$
extends *peds_stopping_to_waiting*
 when

   w2.1e : $peds = stopping$
 then

   w2.1f : $peds := waiting$
 end
Event *cars_waiting_to_moving* $\widehat{=}$
extends *cars_waiting_to_moving*
 when

   w2.2a : $cars = waiting$

        s3.2c : $tl\_cars = \{green\}$
then

        w2.2b : $cars := moving$
end

Event *cars_moving_to_stopping* $\widehat{=}$
extends *cars_moving_to_stopping*
    when

        w2.2c : $cars = moving$
    then

        w2.2d : $cars := stopping$
    end

Event *cars_stopping_to_waiting* $\widehat{=}$
extends *cars_stopping_to_waiting*
    when

        w2.2e : $cars = stopping$
    then

        w2.2f : $cars := waiting$
    end

Event *set_tl_peds_red_to_green* $\widehat{=}$
extends *set_tl_peds_red_to_green*
    when

        s3.1d : $tl\_peds = \{red\}$
        s3.1e : $tl\_cars = \{red\}$
        grd1 : $cars = waiting$
        p3.1a : $peds\_was\_green = FALSE$
    then

        s3.1f : $tl\_peds := \{green\}$
    end

Event *set_tl_peds_green_to_red* $\widehat{=}$
extends *set_tl_peds_green_to_red*
    when

        s3.1g : $tl\_peds = \{green\}$
        grd1 : $cars = waiting$
    then

        s3.1h : $tl\_peds := \{red\}$
        p3.1b : $peds\_was\_green := TRUE$

    end

Event *set_tl_cars_red_to_redyellow* $\widehat{=}$
extends *set_tl_cars_red_to_redyellow*
    when

        grd2 : $peds = waiting$
        s3.2d : $tl\_cars = \{red\}$
        s3.3 : $tl\_peds \neq \{green\}$
        p3.1c : $peds\_was\_green = TRUE$
    then

        s3.2f : $tl\_cars := \{red, yellow\}$
    end

Event *set_tl_cars_redyellow_to_green* $\widehat{=}$
extends *set_tl_cars_redyellow_to_green*
    when

        grd2 : $peds = waiting$
        s3.2g : $tl\_cars = \{red, yellow\}$
        s3.2h : $tl\_peds \neq \{green\}$
    then

        s3.2i : $tl\_cars := \{green\}$
    end

Event *set_tl_cars_green_to_yellow* $\widehat{=}$
extends *set_tl_cars_green_to_yellow*
    when

        grd2 : $peds = waiting$
        s3.2j : $tl\_cars = \{green\}$
    then

        s3.2k : $tl\_cars := \{yellow\}$
    end

Event *set_tl_cars_yellow_to_red* $\widehat{=}$
extends *set_tl_cars_yellow_to_red*
    when

        grd2 : $peds = waiting$
        s3.2l : $tl\_cars = \{yellow\}$
    then

        s3.2m : $tl\_cars := \{red\}$
        p3.1d : $peds\_was\_green := FALSE$
    end
END

B.5   Iteration 4

In this iteration, the push button is introduced into the formal model via refinement.

B.5.1   Artefacts

R2.1	When [peds] are [moving] or [stopping] on the [street], [cars] must be [waiting].
R3.1	If [tl_peds] is [green], then [tl_cars] must be [red].
R3.2	[green] cycles for [tl_cars] and [tl_peds] must alternate.
R4.1	[push]ing a [button] results in [request] to be set, it not yet set.
R4.2'	[tl_cars] may only turn not [green], it a [request] is pending.
R4.2''	If a [request] is pending, [tl_peds] must eventually turn [green].
R4.3	After [tl_peds] turns [red], [request] is reset.
N2.1	The system has additional safety properties.
N1.1	The traffic lights for cars [tl_cars] are usually [green].
N1.2	Between [push]ing the button for the first time in a cycle and [tl_peds] allowing pedestrians to cross, at most [t_1] seconds must pass.
N1.3	Upon turning [green], [tl_peds] must stay green for [t_2] seconds, with a tolerance of 5%.
N4.4	[peds] signal their wish to cross the [street] by [push]ing one of the [button]s.
D3.1	The relationship between [tl_peds] and [peds_signal] is a mapping of [red] to [stop] and [green] to [go], respectively.
D3.2	The relationship between [tl_cars] and [cars_signal] shall adhere to Figure 5.6.
W1.1	Two synchronised traffic lights for cars [tl_cars] are located on the [street], according to Figure 5.1.
W1.2	The state of the car traffic lights is represented by [tl_cars], which represents a subset of [red], [yellow] and [green], meaning that the corresponding light is on.
W1.3	Two synchronised traffic lights for pedestrians [tl_peds] are located on the [street], according to Figure 5.1.
W1.4	The state of the pedestrian traffic lights is represented by [tl_peds], which represents a subset of [red] and [green], meaning that the corresponding light is on.
W1.5	Two buttons are mounted on the bases of the pedestrian traffic lights [tl_peds], according to Figure 5.1, allowing [peds] to [push] them.

W1.6   [peds] can press any of the push [button]s to trigger a [push] event.

R1.3   The traffic light system follows the regulations for traffic lights of Germany (Richtlinien für Signalanlagen, RiLSA)

W2.1   [peds] that are not on the [street] are [waiting]. Upon entering the [street], they are [moving], followed by [stopping], before [waiting] again.

W2.2   [cars] that are not on the [street] are [waiting]. Upon entering the [street], they are [moving], followed by [stopping], before [waiting] again.

W2.3   Upon activating the system, [peds] are [waiting].

W2.4   Upon activating the system, [cars] are [waiting].

W2.5   Conceptually, the traffic lights [peds_ signal] and [cars_ signal] can indicate a [stop] or [go] signal, which is represented in the form of colours.

W2.6   The initial state for [peds_ signal] and [cars_ signal] is [stop]

W2.7   [peds] start [moving] only if [peds_ signal] is [go].   If [peds_ signal] turns to [stop], [peds] that are [moving] are [stopping] and will be [waiting], once they finished crossing.

W2.8   [cars] start [moving] only if [cars_ signal] is [go].   If [cars_ signal] turns to [stop], [cars] that are [moving] are [stopping] and will be [waiting], once they finished crossing.

W2.9   When [peds_ signal] indicates [go], the [cars] are [waiting].

W2.10   When [cars_ signal] indicates [go], the [peds] are [waiting].

S2.1   [tl_ peds] and [tl_ cars] must never be [go] at the same time.

B.5.2   Machine mac04

MACHINE   mac04

REFINES   mac03b

SEES   ctx03

VARIABLES

    peds

    cars

    tl_peds

    tl_cars

    peds_was_green

    request

INVARIANTS

      type_request : $request \in BOOL$

EVENTS

Initialisation

  extended

  begin

      init3.0 : $peds := waiting$

      init3.1 : $cars := waiting$

      init3.2 : $tl\_peds := \{red\}$

      init3.3 : $tl\_cars := \{red\}$

      init : $peds\_was\_green := FALSE$

      init_request : $request := FALSE$

  end

Event *peds_ waiting_ to_ moving* $\widehat{=}$

extends *peds_ waiting_ to_ moving*

  when

      w2.1a : $peds = waiting$

      s3.1c : $tl\_peds = \{green\}$

  then

      w2.1b : $peds := moving$

  end

Event *peds_ moving_ to_ stopping* $\widehat{=}$

extends *peds_ moving_ to_ stopping*

  when

      w2.1c : $peds = moving$

  then

      w2.1d : $peds := stopping$

  end

Event *peds_ stopping_ to_ waiting* $\widehat{=}$

extends *peds_ stopping_ to_ waiting*

  when

      w2.1e : $peds = stopping$

  then

      w2.1f : $peds := waiting$

  end

Event *cars_ waiting_ to_ moving* $\widehat{=}$

extends *cars_ waiting_ to_ moving*

when

        w2.2a : $cars = waiting$

        s3.2c : $tl\_cars = \{green\}$

then

        w2.2b : $cars := moving$

end

Event $cars\_moving\_to\_stopping \; \widehat{=}$

extends $cars\_moving\_to\_stopping$

    when

        w2.2c : $cars = moving$

    then

        w2.2d : $cars := stopping$

    end

Event $cars\_stopping\_to\_waiting \; \widehat{=}$

extends $cars\_stopping\_to\_waiting$

    when

        w2.2e : $cars = stopping$

    then

        w2.2f : $cars := waiting$

    end

Event $set\_tl\_peds\_red\_to\_green \; \widehat{=}$

extends $set\_tl\_peds\_red\_to\_green$

    when

        s3.1d : $tl\_peds = \{red\}$

        s3.1e : $tl\_cars = \{red\}$

        grd1 : $cars = waiting$

        p3.1a : $peds\_was\_green = FALSE$

    then

        s3.1f : $tl\_peds := \{green\}$

    end

Event $set\_tl\_peds\_green\_to\_red \; \widehat{=}$

extends $set\_tl\_peds\_green\_to\_red$

    when

        s3.1g : $tl\_peds = \{green\}$

        grd1 : $cars = waiting$

    then

$\qquad$ s3.1h : $tl\_peds := \{red\}$

$\qquad$ p3.1b : $peds\_was\_green := TRUE$

$\qquad$ s4.3 : $request := FALSE$

$\quad$ **end**

Event *set_ tl_ cars_ red_ to_ redyellow* $\widehat{=}$

extends *set_ tl_ cars_ red_ to_ redyellow*

$\quad$ **when**

$\qquad$ grd2 : $peds = waiting$

$\qquad$ s3.2d : $tl\_cars = \{red\}$

$\qquad$ s3.3 : $tl\_peds \neq \{green\}$

$\qquad$ p3.1c : $peds\_was\_green = TRUE$

$\quad$ **then**

$\qquad$ s3.2f : $tl\_cars := \{red, yellow\}$

$\quad$ **end**

Event *set_ tl_ cars_ redyellow_ to_ green* $\widehat{=}$

extends *set_ tl_ cars_ redyellow_ to_ green*

$\quad$ **when**

$\qquad$ grd2 : $peds = waiting$

$\qquad$ s3.2g : $tl\_cars = \{red, yellow\}$

$\qquad$ s3.2h : $tl\_peds \neq \{green\}$

$\quad$ **then**

$\qquad$ s3.2i : $tl\_cars := \{green\}$

$\quad$ **end**

Event *set_ tl_ cars_ green_ to_ yellow* $\widehat{=}$

extends *set_ tl_ cars_ green_ to_ yellow*

$\quad$ **when**

$\qquad$ grd2 : $peds = waiting$

$\qquad$ s3.2j : $tl\_cars = \{green\}$

$\qquad$ s4.2 : $request = TRUE$

$\quad$ **then**

$\qquad$ s3.2k : $tl\_cars := \{yellow\}$

$\quad$ **end**

Event *set_ tl_ cars_ yellow_ to_ red* $\widehat{=}$

extends *set_ tl_ cars_ yellow_ to_ red*

$\quad$ **when**

$\qquad$ grd2 : $peds = waiting$

    s3.21 : $tl\_cars = \{yellow\}$

  then

    s3.2m : $tl\_cars := \{red\}$
    p3.1d : $peds\_was\_green := FALSE$

  end

Event   *set_request* $\widehat{=}$

  when

    s4.1a : $request = FALSE$

  then

    s4.1b : $request := TRUE$

  end

END

B.6   Iteration 5

This refinement introduces time by adding artificial "ticks" to the formal model.

B.6.1   Artefacts

R1.3	The traffic light system follows the regulations for traffic lights of Germany (Richtlinien für Signalanlagen, RiLSA)
R2.1	When [peds] are [moving] or [stopping] on the [street], [cars] must be [waiting].
R3.1	If [tl_peds] is [green], then [tl_cars] must be [red].
R3.2	[green] cycles for [tl_cars] and [tl_peds] must alternate.
R4.1	[push]ing a [button] results in [request] to be set, it not yet set.
R4.2′	[tl_cars] may only turn not [green], it a [request] is pending.
R4.2″	If a [request] is pending, [tl_peds] must eventually turn [green].
R4.3	After [tl_peds] turns [red], [request] is reset.
R5.1	Between a [request] and [tl_peds] turing [green], at most [t_1] ticks must pass.
R5.2	Upon turning [green], [tl_peds] must stay green for [t_2] ticks.
R5.3	Upon turning [green], [tl_cars] must stay green for [t_3] ticks.
N2.1	The system has additional safety properties.
N1.1	The traffic lights for cars [tl_cars] are usually [green].
N4.4	[peds] signal their wish to cross the [street] by [push]ing one of the [button]s.
N5.1	Each event that modifies traffic light transitions is [1] [tick] long.
N5.2	The length of a [tick] is [1 second] with a tolerance of 5%.
D3.1	The relationship between [tl_peds] and [peds_signal] is a mapping of [red] to [stop] and [green] to [go], respectively.
D3.2	The relationship between [tl_cars] and [cars_signal] shall adhere to Figure 5.6.
W1.1	Two synchronised traffic lights for cars [tl_cars] are located on the [street], according to Figure 5.1.
W1.2	The state of the car traffic lights is represented by [tl_cars], which represents a subset of [red], [yellow] and [green], meaning that the corresponding light is on.
W1.3	Two synchronised traffic lights for pedestrians [tl_peds] are located on the [street], according to Figure 5.1.

W1.4   The state of the pedestrian traffic lights is represented by
        [tl_peds], which represents a subset of [red] and [green], mean-
        ing that the corresponding light is on.

W1.5   Two buttons are mounted on the bases of the pedestrian traffic
        lights [tl_peds], according to Figure 5.1, allowing [peds] to
        [push] them.

W1.6   [peds] can press any of the push [button]s to trigger a [push]
        event.

W2.1   [peds] that are not on the [street] are [waiting]. Upon entering
        the [street], they are [moving], followed by [stopping], before
        [waiting] again.

W2.2   [cars] that are not on the [street] are [waiting]. Upon entering
        the [street], they are [moving], followed by [stopping], before
        [waiting] again.

W2.3   Upon activating the system, [peds] are [waiting].

W2.4   Upon activating the system, [cars] are [waiting].

W2.5   Conceptually, the traffic lights [peds_signal] and [cars_signal]
        can indicate a [stop] or [go] signal, which is represented in the
        form of colours.

W2.6   The initial state for [peds_signal] and [cars_signal] is [stop]

W2.7   [peds] start [moving] only if [peds_signal] is [go].   If
        [peds_signal] turns to [stop], [peds] that are [moving] are [stop-
        ping] and will be [waiting], once they finished crossing.

W2.8   [cars] start [moving] only if [cars_signal] is [go].   If
        [cars_signal] turns to [stop], [cars] that are [moving] are [stop-
        ping] and will be [waiting], once they finished crossing.

W2.9   When [peds_signal] indicates [go], the [cars] are [waiting].

W2.10   When [cars_signal] indicates [go], the [peds] are [waiting].

S2.1   [tl_peds] and [tl_cars] must never be [go] at the same time.

B.6.2   Context ctx05

CONTEXT   ctx05

EXTENDS   ctx03

CONSTANTS

t_1

t_3

AXIOMS

t_1 : $t\_1 = 1$

$$\texttt{t\_3} : t\_3 = 1$$

END

B.6.3 Machine mac05

MACHINE mac05
REFINES mac04
SEES ctx05
VARIABLES

> *peds*
> *cars*
>
> *tl_peds*
> *tl_cars*
> *peds_was_green*
> *request*
> *ticks_peds_green*
> *ticks_cars_green*

INVARIANTS

> ticks1 : $ticks\_peds\_green \in \mathbb{N}$
> ticks2 : $ticks\_cars\_green \in \mathbb{N}$

EVENTS
Initialisation
> *extended*
> **begin**
>
>> init3.0 : $peds := waiting$
>> init3.1 : $cars := waiting$
>> init3.2 : $tl\_peds := \{red\}$
>> init3.3 : $tl\_cars := \{red\}$
>> init : $peds\_was\_green := FALSE$
>> init_request : $request := FALSE$
>> init5.1 : $ticks\_peds\_green := 0$
>> init5.2 : $ticks\_cars\_green := 0$
>
> **end**

Event *peds_waiting_to_moving* $\widehat{=}$
extends *peds_waiting_to_moving*
> **when**
>
>> w2.1a : $peds = waiting$

        s3.1c : $tl\_peds = \{green\}$
        hide_domain : ⊥
then

        w2.1b : $peds := moving$
end

Event $peds\_moving\_to\_stopping \mathrel{\widehat{=}}$
extends $peds\_moving\_to\_stopping$
    when

        w2.1c : $peds = moving$
        hide_domain : ⊥
then

        w2.1d : $peds := stopping$
end

Event $peds\_stopping\_to\_waiting \mathrel{\widehat{=}}$
extends $peds\_stopping\_to\_waiting$
    when

        w2.1e : $peds = stopping$
        hide_domain : ⊥
then

        w2.1f : $peds := waiting$
end

Event $cars\_waiting\_to\_moving \mathrel{\widehat{=}}$
extends $cars\_waiting\_to\_moving$
    when

        w2.2a : $cars = waiting$
        s3.2c : $tl\_cars = \{green\}$
        hide_domain : ⊥
then

        w2.2b : $cars := moving$
end

Event $cars\_moving\_to\_stopping \mathrel{\widehat{=}}$
extends $cars\_moving\_to\_stopping$
    when

        w2.2c : $cars = moving$
        hide_domain : ⊥
then

        w2.2d : $cars := stopping$

        end

Event *cars_stopping_to_waiting* $\widehat{=}$

extends *cars_stopping_to_waiting*

      when

            w2.2e : $cars = stopping$

            hide_domain : $\perp$

      then

            w2.2f : $cars := waiting$

      end

Event *set_tl_peds_red_to_green* $\widehat{=}$

extends *set_tl_peds_red_to_green*

      when

            s3.1d : $tl\_peds = \{red\}$

            s3.1e : $tl\_cars = \{red\}$

            grd1 : $cars = waiting$

            p3.1a : $peds\_was\_green = FALSE$

      then

            s3.1f : $tl\_peds := \{green\}$

            p6.1a : $ticks\_peds\_green := t\_1$

      end

Event *set_tl_peds_green_to_red* $\widehat{=}$

extends *set_tl_peds_green_to_red*

      when

            s3.1g : $tl\_peds = \{green\}$

            grd1 : $cars = waiting$

            p6.1b : $ticks\_peds\_green = 0$

      then

            s3.1h : $tl\_peds := \{red\}$

            p3.1b : $peds\_was\_green := TRUE$

            s4.3 : $request := FALSE$

      end

Event *set_tl_cars_red_to_redyellow* $\widehat{=}$

extends *set_tl_cars_red_to_redyellow*

      when

            grd2 : $peds = waiting$

            s3.2d : $tl\_cars = \{red\}$

            s3.3 : $tl\_peds \neq \{green\}$

$$\text{p3.1c} : peds\_was\_green = TRUE$$

then

$$\text{s3.2f} : tl\_cars := \{red, yellow\}$$

end

Event *set_ tl_ cars_ redyellow_ to_ green* $\widehat{=}$

extends *set_ tl_ cars_ redyellow_ to_ green*

    when

$$\text{grd2} : peds = waiting$$
$$\text{s3.2g} : tl\_cars = \{red, yellow\}$$
$$\text{s3.2h} : tl\_peds \neq \{green\}$$

    then

$$\text{s3.2i} : tl\_cars := \{green\}$$
$$\text{p6.2a} : ticks\_cars\_green := t\_1$$

    end

Event *set_ tl_ cars_ green_ to_ yellow* $\widehat{=}$

extends *set_ tl_ cars_ green_ to_ yellow*

    when

$$\text{grd2} : peds = waiting$$
$$\text{s3.2j} : tl\_cars = \{green\}$$
$$\text{s4.2} : request = TRUE$$
$$\text{p6.2b} : ticks\_cars\_green = 0$$

    then

$$\text{s3.2k} : tl\_cars := \{yellow\}$$

    end

Event *set_ tl_ cars_ yellow_ to_ red* $\widehat{=}$

extends *set_ tl_ cars_ yellow_ to_ red*

    when

$$\text{grd2} : peds = waiting$$
$$\text{s3.2l} : tl\_cars = \{yellow\}$$

    then

$$\text{s3.2m} : tl\_cars := \{red\}$$
$$\text{p3.1d} : peds\_was\_green := FALSE$$

    end

Event *set_ request* $\widehat{=}$

extends *set_ request*

    when

s4.1a : $request = FALSE$

then

s4.1b : $request := TRUE$

end

Event   *tick_peds_green* $\widehat{=}$

   when

p6.1c : $ticks\_peds\_green > 0$
p6.1d : $tl\_peds = \{green\}$

   then

p6.1e : $ticks\_peds\_green := ticks\_peds\_green - 1$

   end

Event   *tick_cars_green* $\widehat{=}$

   when

p6.2c : $ticks\_cars\_green > 0$
p6.2d : $tl\_cars = \{green\}$

   then

p6.1e : $ticks\_cars\_green := ticks\_cars\_green - 1$

   end

END

Bibliography

[Abrial, 2006] Abrial, J.-R. (2006). Formal methods in industry: achievements, problems, future. In *Proceedings of the 28th international conference on Software engineering*, pages 761–768.

[Abrial, 2010] Abrial, J.-R. (2010). *Modeling in Event-B: System and Software Engineering*. Cambridge University Press, 1st edition.

[Abrial et al., 2006] Abrial, J.-R., Butler, M., Hallerstede, S., and Voisin, L. (2006). An open extensible tool environment for Event-B. In *International Conference on Formal Engineering Methods (ICFEM)*, LNCS, New York, NY. Springer-Verlag.

[Abrial et al., 2010] Abrial, J.-R., Butler, M. J., Hallerstede, S., Hoang, T. S., Mehta, F., and Voisin, L. (2010). Rodin: An open toolset for modelling and reasoning in event-B. *STTT*, 12(6):447–466.

[Ambriola and Gervasi, 1997] Ambriola, V. and Gervasi, V. (1997). Processing natural language requirements. In *Automated Software Engineering, 1997. Proceedings., 12th IEEE International Conference*, pages 36–45. IEEE.

[Apt et al., 2009] Apt, K. R., de Boer, F. S., and Olderog, E.-R. (2009). *Verification of Sequential and Concurrent Programs*. Springer, 3rd edition.

[Awad, 2005] Awad, M. (2005). A comparison between agile and traditional software development methodologies. *University of Western Australia*.

[Babar et al., 2007] Babar, A., Tosic, V., and Potter, J. (2007). Aligning the map requirements modelling with the B-method for formal software development. In *Software Engineering Conference, 2007. APSEC 2007. 14th Asia-Pacific*, page 17–24.

[Balduino, 2007] Balduino, R. (2007). Introduction to OpenUP (Open Unified Process). *Eclipse site.*

[Ball, 2008] Ball, E. (2008). *An Incremental Process for the Development of Multi-agent Systems in Event-B.* PhD thesis, University of Southampton.

[Beck, 2001] Beck, K. (2001). *Extreme programming explained: embrace change.* Addison-Wesley.

[Berry, 1999] Berry, D. M. (1999). Formal methods: the very idea – some thoughts about why they work when they work. *Science of computer Programming*, 42(1):11–27.

[Bjørner, 2008] Bjørner, D. (2008). From domain to requirements. In *Concurrency, Graphs and Models: Essays dedicated to Ugo Montanari on the Occasion of his 65th Birthday*, pages 278–300. Springer.

[Brinksma et al., 1995] Brinksma, E., Scollo, G., and Steenbergen, C. (1995). Lotos specifications, their implementations and their tests. In *Conformance testing methodologies and architectures for OSI protocols*, pages 468–479. IEEE Computer Society Press.

[Broy and Rausch, 2005] Broy, M. and Rausch, A. (2005). Das neue V-Modell XT. *Informatik-Spektrum*, 28(3):220–229.

[Budinsky et al., 2009] Budinsky, F., Steinberg, D., Merks, F., and Paternostro, M. (2009). *Eclipse Modeling Framework.* The Eclipse Series. Addison-Wesley Professional, 2nd edition.

[Chung and do Prado Leite, 2009] Chung, L. and do Prado Leite, J. C. S. (2009). On non-functional requirements in software engineering. In Borgida, A., Chaudhri, V. K., Giorgini, P., and Yu, E. S. K., editors, *Conceptual Modeling: Foundations and Applications*, volume 5600 of *LNCS*, pages 363–379. Springer.

[Clarke, 1997] Clarke, E. (1997). Model checking. In *Foundations of software technology and theoretical computer science*, pages 54–56. Springer.

[Clarke and Wing, 1996] Clarke, E. and Wing, J. (1996). Formal methods: State of the art and future directions. *ACM Computing Surveys (CSUR)*, 28(4):626–643.

[Cohn, 2004] Cohn, M. (2004). *User stories applied: For agile software development.* Addison-Wesley Professional.

[Coleman and Jones, 2007] Coleman, J. and Jones, C. (2007). A structural proof of the soundness of rely/guarantee rules. *Journal of Logic and Computation*, 17(4):807.

[Coleman et al., 2005] Coleman, J., Jones, C., Oliver, I., Romanovsky, A., and Troubitsyna, E. (2005). RODIN (rigorous open development environment for complex systems). *EDCC-5, Budapest, Supplementary Volume*, page 23–26.

[Consulting and Ninomiya, 1997] Consulting, M. and Ninomiya, N. (1997). Ariane 5: Who dunnit? *IEEE Software*.

[Cook, 1971] Cook, S. (1971). The complexity of theorem-proving procedures. In *Proceedings of the third annual ACM symposium on Theory of computing*, pages 151–158. ACM.

[Darimont et al., 1997] Darimont, R., Delor, E., Massonet, P., and Lamsweerde, A. v. (1997). GRAIL/KAOS: an environment for goal-driven requirements engineering. In *Proceedings of the 19th international conference on Software engineering*, pages 612–613, Boston, Massachusetts, United States. ACM.

[DeMarco, 1979] DeMarco, T. (1979). *Structured analysis and system specification*. Yourdon Press.

[DEPLOY Project, 2009] DEPLOY Project (2009). Advances in methods (DEPLOY deliverable D15). Technical report, EU-IST "RODIN" Project.

[Edmunds and Butler, 2010] Edmunds, A. and Butler, M. (2010). Tool support for event-b code generation. *WS-TBFM2010*.

[EU FP7 Project, 2012] EU FP7 Project (2008 – 2012). Deploy Project - Industrial deployment of system engineering methods providing high dependability and productivity. http://www.deploy-project.eu/.

[Fabbrini et al., 1998] Fabbrini, F., Fusani, M., Gervasi, V., Gnesi, S., and Ruggieri, S. (1998). Achieving quality in natural language requirements. *Proceedings of the 11 th International Software Quality Week*.

[Fowler and Scott, 2000] Fowler, M. and Scott, K. (2000). *UML distilled: a brief guide to the standard object modeling language*. Addison-Wesley Longman Publishing Co., Inc.

[Goguen and Linde, 1993] Goguen, J. and Linde, C. (1993). Techniques for requirements elicitation. In *Requirements Engineering, 1993., Proceedings of IEEE International Symposium on*, pages 152–164. IEEE.

[Goldin and Berry, 1997] Goldin, L. and Berry, D. (1997). Abstfinder, a prototype natural language text abstraction finder for use in requirements elicitation. *Automated Software Engineering*, 4(4):375–412.

[Gotel and Finkelstein, 1994] Gotel, O. and Finkelstein, A. (1994). An analysis of the requirements traceability problem. In *Proceedings of the First International Conference on Requirements Engineering*, page 94–101.

[Gross and Yu, 2001] Gross, D. and Yu, E. (2001). From non-functional requirements to design through patterns. *Requirements Engineering*, 6(1):18–36.

[Gunter et al., 2000] Gunter, C. A., Jackson, M., Gunter, E. L., and Zave, P. (2000). A reference model for requirements and specifications. *IEEE Software*, 17:37–43.

[Guttag et al., 1993] Guttag, J., Horning, J., Garl, W., Jones, K., Modet, A., and Wing, J. (1993). Larch: languages and tools for formal specification. In *Texts and Monographs in Computer Science*. Citeseer.

[Hall et al., 2002] Hall, J. G., Jackson, M., Laney, R. C., Nuseibeh, B., and Rapanotti, L. (2002). Relating software requirements and architectures using Problem Frames. In *Requirements Engineering, 2002. Proceedings. IEEE Joint International Conference on*, page 137–144.

[Hallerstede et al., 2012] Hallerstede, S., Jastram, M., and Ladenberger, L. (2012). A method and tool for tracing requirements into specifications. To be published in Electronic Communications of the EASST.

[Hammad et al., 2009] Hammad, M., Collard, M. L., and Maletic, J. I. (2009). Automatically identifying changes that impact Code-to-Design traceability. *ICPC*.

[Hansson et al., 2012] Hansson, D. et al. (2012). Ruby on Rails. *Website: http://www.rubyonrails.org*.

[Hoare, 1978] Hoare, C. (1978). Communicating sequential processes. *Communications of the ACM*, 21(8):666–677.

[Hoare, 2004] Hoare, C. (2004). *Communicating Sequential Processes*. Prentice Hall International.

[Hoare and Jifeng, 1998] Hoare, C. A. R. and Jifeng, H. (1998). *Unifying Theories of Programming*. Prentice Hall.

[Hood et al., 2007] Hood, C., Mühlbauer, S., Rupp, C., and Versteegen, G. (2007). *IX-Studie Anforderungsmanagement*. Heise-Zeitschr.-Verl.

[Hood and Wiebel, 2005] Hood, C. and Wiebel, R. (2005). *Optimieren von Requirements Management & Engineering: mit dem HOOD Capability Model*. Springer.

[IEEE, 1997] IEEE (1997). Recommended practice for software requirements specifications. Technical Report IEEE Std 830-1998, IEEE.

[IEEE, 2010] IEEE (2010). Systems and software engineering – vocabulary. Technical Report ISO/IEC/IEEE24765, IEEE.

[IIBA, 2009] IIBA (2009). *A Guide to the Business Analysis Body of Knowledge*. International Institute of Business Analysis, 2nd edition.

[Jackson, 2001] Jackson, M. (2001). *Problem Frames: analysing and structuring software development problems*. Addison-Wesley/ACM Press, Harlow England New York.

[Jastram, 2010] Jastram, M. (2010). ProR, an open source platform for requirements engineering based on RIF. *SEISCONF*.

[Jastram, 2011] Jastram, M. (2011). ProR - Eine Softwareplattform für Requirements Engineering. *Softwaretechnik-Trends*, 31(1).

[Jastram, 2012a] Jastram, M. (2012a). Strukturierung von Anforderungen für eine enge Integration mit Modellen. *ReConf*.

[Jastram, 2012b] Jastram, M. (2012b). Using the Eclipse Requirements Modeling Framework. In Maalej, W. and Thurimella, A. K., editors, *Managing Requirements Knowledge*, chapter 16. Springer.

[Jastram and Brökens, 2012] Jastram, M. and Brökens, M. (2012). ReqIF in der Open Source: Das Eclipse Requirements Modeling Framework (RMF). *ReConf*.

[Jastram and Ebert, 2012] Jastram, M. and Ebert, C. (2012). ReqIF: Seamless requirements interchange format between business partners. To be published in IEEE Software.

[Jastram and Graf, 2011a] Jastram, M. and Graf, A. (2011a). ProR, eine auf RIF/ReqIF basierende Open Source Plattform zum Anforderungsmanagement. *ReConf*.

[Jastram and Graf, 2011b] Jastram, M. and Graf, A. (2011b). Requirement traceability in Topcased with the requirements interchange format (RIF/ReqIF). *First Topcased Days Toulouse*.

[Jastram and Graf, 2011c] Jastram, M. and Graf, A. (2011c). Requirements Modeling Framework. *Eclipse Magazin*, 6.11.

[Jastram and Graf, 2011d] Jastram, M. and Graf, A. (2011d). Requirements, traceability and DSLs in Eclipse with the requirements interchange format (RIF/ReqIF). Technical report, Dagstuhl-Workshop MBEES 2011: Modellbasierte Entwicklung eingebetteter Systeme.

[Jastram and Graf, 2012] Jastram, M. and Graf, A. (2012). ReqIF – the new requirements standard and its open source implementation Eclipse RMF. Technical report, Commercial Vehicle Technology Symposium.

[Jastram et al., 2011] Jastram, M., Hallerstede, S., and Ladenberger, L. (2011). Mixing formal and informal model elements for tracing requirements. In *Automated Verification of Critical Systems (AVoCS)*.

[Jastram et al., 2010] Jastram, M., Hallerstede, S., Leuschel, M., and Russo Jr, A. (2010). An approach of requirements tracing in formal refinement. In *VSTTE*. Springer.

[Jones, 1990] Jones, C. (1990). *Systematic software development using VDM*, volume 103. Prentice-Hall.

[Jones et al., 2007] Jones, C. B., Hayes, I. J., and Jackson, M. A. (2007). Deriving specifications for systems that are connected to the physical world. *Lecture Notes in Computer Science*, 4700:364.

[K. Forsberg and Cotterman, 2005] K. Forsberg, H. M. and Cotterman, H. (2005). *Visualizing Project Management*. John Wiley and Sons, 3 edition.

[Kaindl, 1997] Kaindl, H. (1997). A practical approach to combining requirements definition and object-oriented analysis. *Annals of Software Engineering*, 3(1):319–343.

[Kang and Jackson, 2010] Kang, E. and Jackson, D. (2010). Dependability arguments with trusted bases. In *Requirements Engineering Conference (RE), 2010 18th IEEE International*, page 262–271.

[Kennedy, 1961] Kennedy, J. F. (1961). Special joint session of congress.

[Kovitz, 1998] Kovitz, B. (1998). *Practical software requirements: a manual of content and style.* Manning Publications Co.

[Kruchten, 2004] Kruchten, P. (2004). *The Rational Unified Process: An Introduction.* Addison-Wesley Professional.

[Leuschel and Butler, 2003] Leuschel, M. and Butler, M. (2003). ProB: A model checker for B. *FME 2003: Formal Methods,* pages 855–874.

[Loesch et al., 2010] Loesch, F., Gmehlich, R., Grau, K., Jones, C., and Mazzara, M. (2010). Report on pilot deployment in automotive sector (D19). Technical Report D7, EU-IST "RODIN" Project.

[Marincic et al., 2007] Marincic, J., Wupper, H., Mader, A., and Wieringa, R. (2007). Obtaining formal models through non-monotonic refinement. Technical report, Centre for Telematics and Information Technology University of Twente.

[Matoussi et al., 2008] Matoussi, A., Gervais, F., and Laleau, R. (2008). A first attempt to express KAOS refinement patterns with Event B. In *Proc. of the Int. Conf. on ASM, B and Z (ABZ). Lecture Notes in Computer Science, Springer-Verlag,* page 12–14.

[McConnell, 2004] McConnell, S. (2004). *Code complete: a practical handbook of software construction.* Microsoft press.

[Moller and Tofts, 1990] Moller, F. and Tofts, C. (1990). A temporal calculus of communicating systems. *CONCUR'90 Theories of Concurrency: Unification and Extension,* pages 401–415.

[Nielsen et al., 1989] Nielsen, M., Havelund, K., Wagner, K., and George, C. (1989). The raise language, method and tools. *Formal Aspects of Computing,* 1(1):85–114.

[OMG, 2011] OMG (2011). Requirements interchange format (ReqIF) 1.0.1. http://www.omg.org/spec/ReqIF/.

[Parnas and Madey, 1995] Parnas, D. L. and Madey, J. (1995). Functional documents for computer systems. *Science of Computer programming,* 25(1):41–61.

[Plagge and Leuschel, 2010] Plagge, D. and Leuschel, M. (2010). Seven at one stroke: LTL model checking for high-level specifications in B, Z, CSP, and more. *International Journal on Software Tools for Technology Transfer (STTT),* 12(1):9–21.

[Pohl, 2007] Pohl, K. (2007). *Requirements Engineering. Grundlagen, Prinzipien, Techniken.* Dpunkt.Verlag GmbH, 1 edition.

[Praxis, 2003] Praxis (2003). Reveal – a keystone of modern systems engineering.

[Project Management Institute, 2008] Project Management Institute (2008). *A Guide to the Project Management Body of Knowledge:.* Project Management Institute, 4 original edition.

[Royce, 1970] Royce, W. (1970). Managing the development of large software systems. In *Proceedings of IEEE WESCON*, volume 26, page 1–9.

[Rupp, 2007] Rupp, C. (2007). *Requirements-Engineering und - Management: professionelle, iterative Anforderungsanalyse für die Praxis.* Hanser, München [u.a.], 4., aktualisierte und erw. aufl. edition.

[Schneider, 2001] Schneider, S. (2001). *The B-method: an introduction.* Palgrave Macmillan, Basingstoke.

[Schwaber, 2004] Schwaber, K. (2004). *Agile project management with Scrum*, volume 7. Microsoft Press Redmond (Washington).

[Sippel et al., 2008] Sippel, H., Jastram, M., and Bendisposto, J. (2008). *Die Eclipse Rich Client Platform: Entwicklung von erweiterbaren Anwendungen mit RCP.* Software und Support Verlag.

[Snook and Butler, 2006] Snook, C. and Butler, M. (2006). UML-B: formal modeling and design aided by UML. *ACM Trans. Softw. Eng. Methodol.*, 15(1):92–122.

[Tennant, 2005] Tennant, N. (2005). Relevance in Reasoning. In Shapiro, S., editor, *The Oxford Handbook of Philosophy of Mathematics and Logic*, chapter 23, pages 696–726. Oxford University Press.

[Van Lamsweerde et al., 2001] Van Lamsweerde, A. et al. (2001). Goal-oriented requirements engineering: A guided tour. In *Proceedings of the 5th IEEE International Symposium on Requirements Engineering*, volume 249, page 263.

[Wiegers, 2003] Wiegers, K. (2003). *Software Requirements: Practical Techniques for Gathering and Managing Requirements throughout the Product Development Cycle.* Microsoft Press, Redmond Wash., 2nd edition.

[Wing, 1990] Wing, J. (1990). A specifier's introduction to formal methods. *Computer*, 23(9):8–10.

[Woodcock and Davies, 1996] Woodcock, J. and Davies, J. (1996). *Using Z: specification, refinement, and proof*, volume 39. Prentice Hall.

[Yu, 1997] Yu, E. (1997). Towards modeling and reasoning support for early-phase requirements engineering. *Requirements Engineering*, page 226.

[Zave, 1997] Zave, P. (1997). Classification of research efforts in requirements engineering. *ACM Computing Surveys (CSUR)*, 29(4):315–321.